2012 and the Ring of Light

WHEN MANKIND FINALLY GROWS UP

by

Nancy E. Shaffron

Edited by Patricia Kaspar

authorHOUSE®

AuthorHouse™
1663 Liberty Drive, Suite 200
Bloomington, IN 47403
www.authorhouse.com
Phone: 1-800-839-8640

First published by AuthorHouse 8/20/2007

ISBN: 978-1-4343-2102-2 (e)
ISBN: 978-1-4343-2101-5 (sc)

Printed in the United States of America
Bloomington, Indiana

This book is printed on acid-free paper.

Cover photo courtesy of NASA, 89-HC-220.jpg

CONTENTS

PREFACE

In the year 2012 life on Earth as we know it will be changed forever. Is it divine intervention on a blind society? Or is it simply Gaia responding to physical Earth changes brought on by both Man and Earth movements set in motion eons ago?

Either way, the arrival of the year 2012 marks the transition of Mother Earth from the end of one celestial cycle around her Sun to a new cycle. With that transition will come a cycle of change of epic proportions in all areas of our earthbound existence.

Will we be ready? Can we be ready? If this is our last chance to 'get it right,' something has to change now. Some say quiet change through guidance from outside has been occurring for a long time. Others feel only a dim angst pushed by a force persistent but not recognized, ignoring the glaring areas of change which must occur if we are to continue life here on Earth.

This is the story of a participation and an awareness in which humankind, one last time, is given the opportunity to live.

"On which side are our gods?" we ask. *"Ours?"*

I — Arrival

Two thousand years ago, unseen in the vast blue sky and suspended in a shining spaceport held by universal powers, celestial governors prepared to perform their roles, the readying of Earth and her people for survival. In this shiny spot in the heavens, galactic entities OmRa and Theisha observed the tiny blue orb below. From their vantage point they would prepare earthlings for survival in the coming catastrophic world change.

OmRa looked down at the primitive beings gathering quietly at the base of a craggy hill. "We must provide the necessary genetic modifications if the people of Earth are to take responsibility for their own survival. We have much to do."

"Most of the fourth race of humankind will not survive the transition," mused Theisha, "but its replacement will be by an evolved human race, the planned and foreseen fifth and final race of people on Earth, as those who live for tomorrow."

"It will require centuries in Earth years for the changes to evolve. Yet more rapidly than they can comprehend, their year 2012 approaches."

"This tiny planet has much to offer. We must work to bring it into harmony with our universe."

"Mahrianne will be our emissary. This time she must succeed in shepherding the changes through the generations."

The sound of her name swept down through time, a faint blush of her eerie beauty once again alive on the Blue Planet. Mahrianne comes! The sound blew past the ancient rocky tor, across fields and whistled around farmhouses as it had through time. Within the small dwellings, around the evening fires, the Old Ones looked into one another's eyes, then into the flames. A small curl of smile played about the corners of wrinkled mouths, a sharing of ancient knowledge and of good things to come. Women nervously gathered children. Men quieted the dogs. Within each house all

1

drew nearer the fire to wait, breathing whispers of her name, "Mahrianne, Mahrianne," softly, softly, reverently.

Outside, a nearby hill served as a magnet drawing eerie flashes of light to its jagged crest, silhouetting it as an aura of things to come, harbingers of fear and change. Apprehension spurred from deep within a greater need, from some cosmic backdrop unknowable to the frightened creatures huddled in their dwellings of stone and skins, silent, waiting, watching, knowing things not within the boundaries of their lives. Shades of omens foretold and handed down through time—remnants of things known before the ice and all but forgotten, things from other eras, now lost in myth, reinforced or embellished by the Sayers, generation to generation.

For several evenings the local country folk had seen flashes of light playing about the top of the nearby hill. Its crest, piercing thrusts of rock, appeared now as figures, now as animals. At dusk the quiet settled. The village folk had felt the change, all were waiting, babes fed and bedded. Of the adults and youths, none slept but waited silently, some primitive knowing of change to come. Only the Druids stirred and made ready to bear witness as the Guardians and Watchers. Clothed in their coarse robes, feet secure in heavy skins, they made their way to a place close by the mounds surrounding the hill. And facing east they awaited the events to which they would bear witness.

As darkness fell the full Moon began to rise casting its secret light on the landscape. No breeze stirred. Silent, all waited. Faintly at first, thin strands of light zigzagged across a breach on the horizon knitting dusk to dark. A light crackling of something unknown began to stir just at the breach between Earth and sky, growing quietly, invading the light play. The air stirred about the waiting figures pulling their attention toward a mound crowned by an ancient craggy yew tree, roots gripping what hold they could on the rocky hillock.

Around the aged tree and rocks hardy grasses had crept up the mound leaving only room for a blazing fire to spread its warmth. Tortured brush grew reluctantly amongst the boulders competing for positions with the rocks protruding from the moss and lichen-covered earth. There, in that one barren space free of rock or plant, guarded by the ancient tree, a ball of crackling light, effervescent in blues and yellows, swirled as it took form, an apparition—a female visage sat cross-legged amongst the time-weathered rocks on the low hill.

Jagged streaks of light now crisscrossed between space and clouds. As the full Moon rose bright and somber to supervise the play of the winds, all quieted, waiting for something, what? A sense of knowing emanated from some place deep in the human psyche, awesome and binding to the inhabitants of that rocky coast of England. The long wait had ended and change would come, slowly at first, barely noticed. This they knew.

Their eyes turned to the crone sitting alone. Her face was the face of time itself, a map of a thousand journeys through time and space, bearer of change, conduit to the Blue Planet. Too many cares had carved the lines accenting her features, etching her skin into a map of fine lines.

Quietly, the eldest Druid carried a small wooden bowl containing a deerskin pouch to the crone and placed it gently in her withered hands. The stones in the pouch, each carved with a different glyph, were runes from an earlier time, the glyphs an alphabet for an unspoken language of magical powers.

Eyes closed, the crone sat, tattered skirts layered over knees bent by more than the position they now assumed. Large stones, larger by far than the objects held in the claw-like hands, lay strewn about the hillock upon which she sat, stones laid long before by Druids in performance of some ancient rite of passage.

Moon shadows played against the visage on the hillock, now dancing, now frozen, shy, bold, eerie. A sound as elusive as the Moon shadows trilled through the air, rose and fell as if the stones themselves, or the shadows of dancers from antiquity, breathed a breath of meaning, somewhere else, in some other time.

The ancient hand scooped the pouch of rocks from the bowl rattling them with imported meaning. Eyes glowing with some inner purpose, emanating color not of this Earth but fueled by some ancient source, lost in myths, transported through time, she plucked objects from the pouch one at a time savoring each in her palm. Bony fingers, capped by broken and dirty nails, belied the strength that underlay their dried and thin skin. A shriveled parched smile bared crooked yellow teeth, fossils of time, tales of toil, glaring, laughing, crying.

As each object found its way into her time-weathered hand, she held it for a moment, slowly chanting words guttural and resonant, a constant stream of sound, loud, haunting, and magnified by the stones scattered around that beat against the song of the shadows. The Moon rose higher in the night sky. If the crone noticed it or the cold of the air there was no sign. Object by object was withdrawn from the pouch, pressed to her chest and chanted, no chant like any other, just as no one of the objects was exactly like any other. Each in turn was passed from her bony chest to the lined lips above, then placed in a shape on the ground in front of her, one rune for each of the six positions of the runic cross.

Drawing strength from its watchful silence, she sat among the roots of the ancient yew believed by these people to have sprouted from some primordial source or seed. Chanting softly she drew the runes, one at a time, exactly the way they came from the pouch, her voice an incantation between one space and time and that of another space and time.

She lifted the first stone, its plain surface shiny in the firelight. Soft sounds, words perhaps, floated as mist among the viewers. *The past*, they thought they heard. *The rune of destiny for all previous existences, the link to the fate of the castor of the rune.* She placed the stone onto the earth.

The second stone was raised, the Kano glyph. *The present!* Just a whisper. *The rune of fire, the end of darkness in the castor's life, the sign of passion and sensuality.*

The air chilled. The watchers huddled closer. Ancient hands lifted the third stone, the Dagaz glyph. *The future!* Her words hung in the air. *A massive turnover in life, a new life with all darkness gone and the Sun to light the way.* With the placement of the stone a runic cross began to form at her feet.

The Raido glyph next, words floating slowly down upon them. *Now is the moment to act. Follow the heart. There will be union and reunion and the end of conflict.*

She lifted the fifth stone, the Berkana glyph. *This is the challenge.* Louder now, as in a chant. *There will be fertility, the birth of the new humans. You must progress carefully.*

The sixth and last, the Gebo glyph. *The rune of freedom, individuality and self-confidence, the achievement of the desired results.* Stronger now and resolute. *And for the caster of the stones, choice!* As Gebo was placed into the runic cross, the air about Mahrianne moved. The faint brushing sound slowly became a voice beckoning from nowhere, calling a name too faint to catch.

The Druids gathered round, watchful, ever changing—chimeras in their heavy shrouds, hovering about the ancient ceremony, guiding its outcome. As each rune was cast, its worn surface gleamed, lit by a flash of lighting, its carved symbol made clear in the eerie light. The crash of the accompanying thunder rattled the earth, liquid under the watchers' feet, undulating with the grasses moved by an unseen force. Flames danced casting shadows into the night. A log tumbling from the fire sent swarms of sparks up into the shadows, legions of fireflies and fairies set free to pleasure themselves in their flight.

Eerie shadows stretched and shrank, bobbed and danced, distorted semblance of all fears, ancient rites of Druids and sorcerers, mockeries of rituals performed on wild moonlit nights, of blood and change and the unknowable to come. Witnessing eyes had peered through cracks in stonewalls or spaces in rough shutters. Children were hidden. Offerings were made. On quieter nights these ancient peoples had waited, keeping the secret, to greet their new and changed bairn. Passing centuries had not altered these secrets and yet the changes had come even as humankind spread itself inexorably across the land.

Then came the silence. No sound other than a few insects, a few small animals scurrying about—all quiet now. Mahrianne stood in the center of the circle of stones, small against their mass, her hair flowing about her head, liquid fire covering her shoulders, caressing her breasts. With each rune thus blessed, her visage had altered, traces of time slowly vanishing. She picked up her stones and returned them to their pouch that she then slipped into a pocket guarded by an unseen opening in her garment. Together they waited, ready to perform the work of time, the improvement of the human genome.

A strong male wind, taking her as she transformed, shaped her to his whim. Lightning crackled, came and went, commanded by the ethereal wind. Tangled strands of long wild hair, blown by the frenzied wind that involved only her, began to change. White became fiery auburn, red hued. Snarls of great age smoothed into flowing waves. The night wind brushed her hair back over rosy cheeks gracing high cheekbones. Soft deep brown eyes smiled as lush hair tumbled down over the now firm body. A soft sigh passed through her ripe full-lipped mouth.

Now shaping her neck, the wind toyed with the high firm breasts and gently, as in a lover's kiss, brushed the hard nipples, erect in the cold air. The wind stripped the tattered filthy robes from her form as it took its final shape, sensual, youthful yet never losing the wisdom of the past eons. A soft nondescript garment now covered her form, flowing softly in a motion of its own. Sandals of ancient design now guarded her feet.

The knowing country folk had waited, waited to know her again as she had been known many times in the past, waited to know her, secure in their deep and secret knowing of her duty to be performed, the awakening of unused DNA segments built into the human genome, DNA awaiting its pre-ordained awakening. They watched, flesh shivering with the power of the forces at play, waiting for the lush body to be clothed anew, watching the mass of red hair, long waves of molten fire flowing from her proud head. They held their breath until they heard the laugh of exhilaration as young blood coursed through sinew and muscles over strong bones.

Her hair twisted wildly each time the sky was rent by the strange lightning. Iridescent green and orange flowed down the white streaks of violence that tore at the night sky. Each strike returned blood red, streaked by deep purple and met with the sky as its thunderous crash terrorized the silent watchers and those transfixed in the portals of their stone huts. These, both terrified and awestruck, were linked irretrievably with ancient myths of gods and demons that visited Earth at their will, always leaving marks and changes, sometimes good, sometimes not. They peered in stony silence, waiting shivering in the chill of the storm and clinging to each other fearfully, awed by the unearthly display of power.

Then, hands outstretched she tossed her head as a wild mare running. Hands caressed flesh, feeling the strength of her body, the rush of her loins, her laugh ringing across the wind-swept hills, echoing from the circle of stones. No soul or heart hearing that sound would ever feel less than a god itself. Nor would it ever be discussed save for tales of creation around quiet fires enclosed in the stone dwellings in the south of England. This she would be at her whim, ephemeral otherwise, for two thousand more years.

Mahrianne picked up the wooden flute, savoring its skin, wood worn smooth over time, touched by hands human and ethereal. Her lips quavered slightly at its touch, warm, yet cool and soothing, a lover long forgotten. Her dark eyes closed as her fingertips found the openings in the pipe of the flute, drifting back in time to woods such as these, woods sheltering rites as primitive and sensual as these. Slowly she filled her being with breath, and as she gently exhaled its sweetness, the eerie haunting notes of an ancient melody floated over the tall grasses brushing new flowers and caressing high boughs in their fulfillment.

As she played, a faint blue mist formed and grew in the notes and moved forward toward two who lay twined in their ecstasy. Two disembodied forms slowly took shape, drifting with the haunting music, magic to the two. Moonlight threaded through the trees barely reaching the small clearing. Soft grasses and tiny flowers carpeted the warm ground, touched ever so slightly by cool breezes from somewhere in the trees. The changes had begun.

Mahrianne was here! Keeper of the Secrets, Light Worker of change. Mahrianne, shepherd of the final four children, bearers of the links to sacred tablets, key to the future of the Blue Planet, third from the Sun, galaxy of the World of Light. For the next two thousand plus years she would assist those Light Beings, overseers in helping the Blue Planet achieve its place in the World of Light.

The baying hounds looked to the Moon now illuminating the eerie scene and the strange acts taking place, connected at some deep seat in their dark primitive brain stems. Understanding more than they could express, they became a chorus in synchronization with powers far beyond those manifest at the surface of this round globe, on this rolling terrain, in this deep moonlit night. The waiting ended, *anima mundi*—the organizing essence of the physical universe, spirit essence of physical nature thus enjoined.

As the morning light came, Mahrianne drew from the assembled Druids the earthly Guardians to assist her work. They now stood in the gold rays of the bright morning sunlight gracing the rocky hillock. She spoke to them at length in quiet tones, often mind to mind. By the time she completed her instructions the Sun was high in the sky. She held each with her mind for a long moment. There was much work to be done, beginning with a session with the overseers, those who would bear the responsibility to complete the planned changes. There could be no wasted time.

II – THE DANCE BEGINS

For the past 20 centuries, almost no one had noticed the faint shining spot in the northern sky. It had always been there, they thought, these people of Earth. It was probably just the way Earth's magnetic field was. "You know," they mused, "with the Aurora Borealis and all. No need to speculate on something that probably isn't even there." Life went on and no scientist commented on it. In actuality it had been there a very long time in Earth cycles. However, since it was apparently innocuous and had no unusual characteristics except that it was there, it remained unnoticed.

"It is just as well that they do not notice," said OmRa. "The earthlings could not have known of our mission to them from the Galaxy."

"Yes," replied Thiesha. "It is time to manage one last round with the diverse human beings inhabiting Earth. Three times Earth-driven catastrophes have all but wiped them out. Each time they have come back stronger and more advanced. Now the currant inhabitants will face the fourth and final battle with their planet Earth. If they survive, this struggle would make them the fifth, and final, race to inhabit Earth."

"It is strange that this sequence of trials has been carried in the myths of all races on the planet."

"True. Myths regarding such catastrophes are similar race by race, tribulations brought on by their own misdeeds, survivors chosen by some unknown power. Each rejuvenation of the human race has been accompanied by a set of reasons for who was chosen to survive, how they survived, and how they learned what they had done wrong, or not done right, to cause such disruption as had occurred—then and now. The question is, have they learned? I believe Mahrianne has done well. Now we shall see."

For the better part of the past two thousand years, Mahrianne and these Light Workers, Guardians of the new humans, had undertaken the task of developing the roots of a new race. This fifth and final race of humans would now have the opportunity to prove itself worthy of joining the World of Light. Led by Mahrianne, the Light Workers had solidly moved to select candidates for the complex DNA enhancements. Marching through generations of promising candidates, often thwarted by earthly mishaps, they had watched over carefully generated strains of humans until numerous viable strains with the necessary qualities had been cultivated.

As time wore on, the web of connected and enlightened souls grew, watched without interference from the waiting Overseers, their appearance only a small shiny spot in a remote segment of the heavens—if one really bothered to look. Mahrianne knew her task. The level of genetic modification was sufficient. It was time for processing the chosen four. No mistakes could be made.

Strange new DNA strands blinked in and out, iridescent quasars of evolution, each more intricate than the previous, bearing its own secret message of change. Twisting threads of pastels, resplendent in their garb of colors indescribable in their beauty and unseen by man.

Weaving gently, Mahrianne watched the couple, graceful in her dance of joining, a dance performed repeatedly over eons of time, always the same yet more intricate and involved with the introduction of each new alteration to the human genome. The intricate moves performed in the dance reflected images evoked by obscure and exquisite holograms of subtle alterations to be revealed as the new humans matured and executed their role in man's evolution.

Those entities also watching, carefully performing their roles in the unfolding creation of the harbingers of a new age, witnessed a light show both powerful and delicate, beyond technology and directed through their presence.

As this dance of creation began, faint trilling sounds accompanied by the eerie tinkle of tiny bells wove into a web of sound around the unearthly gathering. Weaving their magical blend of rhythm, the vibrating sound of distant didgeridoos came, more felt than heard, resonating through the collective mind of the participants. Imperceptibly, the resonance invaded both mind and body, removing any thoughts other than those of the moment, replacing them with the pure energy of knowing. As their human players neared the gathering, the sounds of the ancient instruments swelled and pulsed, vibrating at resonant frequencies that penetrated the souls of the watchers as their complex, driving rhythms knit the consciousness of the unearthly group into one collective consciousness.

Chimes of heavenly tone joined the flow, punctuated by a new throbbing, pulsing beat of drums of many proportions and timbre as other humans emerged from the surrounding forest. Human players of frightening visage joined the play, backdrop to creation of a New Age.

Afloat over the spell of this web of sound, flutes of wind painted haunting, crying sounds that cut to the very essence of each present, painted people of ancient origins, watchers and humans, ghostly iridescent light beings and the oblivious couple.

III — DR. JASON TOWNE

"This is not yet a family, Thiesha. It is children who must bring this about."

"Patience, OmRa. Mahrianne has chosen well. The players do not yet know that this is the beginning.

Jason Towne grabbed his bags and hurried through the gate absently searching the waiting crowd for his friend and colleague. The Africa-to-Los Angeles flight had been impossibly draining, but not as draining as the last few weeks. If only . . . He'd said that to himself a thousand times. If only he could have said goodbye. Not that it would have mattered. His father hadn't spoken to him in the two years since he'd left home.

"Hey, Jason! Over here!" The shout bellowed out across the crowd streaming out of customs with their belongings and scurrying to get out of the airport. Greg Sutherland's sun-baked face broke into a broad smile as he nudged and pushed his way toward his partner and friend.

Jason squeezed out a tired smile. "Glad to see you Greg. Thanks for picking me up!"

"You look beat, old buddy." Greg took a metal container from Jason's arm and left him with his carry-on bag. One look at his friend's face, however, threw a chill over the excitement at seeing his friend again. He knew Jason well. They'd shared work on the Institute's hush-hush DNA research experiment, Jason as the core-classified DNA-specific expert and Greg as the psychologist and geology expert.

"What's the frown for?" asked Greg.

"Just thinking."

"Don't do that, man! You'll hurt your head!"

The hackneyed joke broke the tension. "What's up, anyway?" When no response came, Greg waited. His friend was usually deep in thought or perhaps worried about something.

Jason took him by the arm steering him away from the crowd. "Let's get to the car first. I have that cold spot in my mind's eye again . . . Hey, careful! Don't drop that container!"

"I won't! What's so special, anyway?"

"My Dad left it to me in his will. There's some sort of artifact in it." He paused and looked away. "It's the only thing he ever left me other than that old desk of his in my office."

Greg heard the wistfulness and tried to lighten the moment. "Well, what'd he leave your brother?"

Jason paused, then said, "Everything else."

Oh boy, thought Greg. *So that's it!* "Your dad must have been upset at something."

"Yeah. I guess he was still miffed. After all of my schooling he thought at least I could do my research in Africa. He never wanted me to study DNA. I think he was afraid of where the results could lead."

"Well, your chances for advancement are certainly better here at the Institute. If your mom were still alive I'll bet she would have looked at it differently."

"It doesn't matter now."

They threw the bags in the car and climbed in. "What kind of artifact's in that container, anyway?" asked Greg.

"I don't know. I haven't opened it yet. I really don't care if I have that thing but Dad's will specified that I take it. There's a drawer in the desk someplace where it belongs. At least I'll have something to remember him by."

Greg tried again to cheer him up. "Yeah, sure Jason. Here we go again. Last time you went to see your folks you came back with an old desk that now takes up most of your office. This time you bring back something to put in it."

Jason grimaced, then laughed. "The desk's been in the family for years. I never thought much about it." But he had thought about it a great deal. It was a big hardwood desk with lots of drawers and secret compartments, one of which was probably meant for that box. He thanked Greg again for picking him up and turned his thoughts to home. "How's the Civil Defense effort going?"

"Great! Our work's beginning to pay off. We should've started even sooner with all the talk about our long-overdue earthquake. At least we've got some good people on the team. We'll meet next week. I'll bring you up to date then."

Back at his apartment Jason finished some lukewarm pizza and washed it down slowly with a cold Sam Adams he'd picked up at the grocery store. Absently, he dropped the empty bottle into a wastebasket along with his round trip ticket stubs and gave it a kick, less in anger than in sadness. He was settled now here in Los Angeles, proud of his job and proud of his new status with a doctorate in an important new field of research. Two years of his widely recognized discoveries at the forefront of DNA research hadn't gained his father's respect nor softened his hard-nosed attitude. Jason had planned eventually to go back to his old home in Johannesburg and settle things there, sort out what to ship and what to give away, but he'd expected it to be under more favorable circumstances.

During the past few weeks he'd spent time alone with family members and others who'd helped him reach his goals. He'd also said goodbye to old friends and set up some communication links. Someday he'd planned to tell his dad how much he loved and respected him and also thank him for helping him through his long education. *Too late now to prove myself in his eyes,* thought Jason sadly. *Maybe someday I'll make a difference somewhere.*

Dr. Jason Michael Towne had been sitting in his lab for hours going over the paper he'd been reading, one of a stack of papers on the molecular mechanisms of gene regulation. God, the sequencing lab was hot! It wasn't, but the old salvaged school clock on the wall said he'd been sitting there four hours and ten minutes. Jason rubbed his burning eyes. Five forty, he thought. While reading a thought had briefly flickered through his brain, danced about and escaped, yet it had left him with the off-center sensation that he was missing something important.

Jason's interest in genetic engineering had begun by accident when it became a required course for a PhD in medical research. At the University of Johannesburg in South Africa some quirk of fate assigned him to a course in molecular biology rather than another he had selected. After providence cast him into genetic research, life became exciting. At the same time, his choice drove the final wedge between him and his father. He'd plunged totally into the field of molecular genetics. His contributions to understanding the mechanisms of change in genetic material had earned him not only the PhD he sought but also the financial support needed to pursue his interests in molecular biological research. The field of gene expression thus became his life.

North Western Research Institute, a newly formed company in California, was one of the least visible, most technically advanced research labs in the U.S., just the way they wanted it. No stock. No commercials. No bothersome meddling press or investors, just two plain unpretentious brick buildings connected by a closed-in breezeway, and two huge underground state-of-the-art labs shielded from electro-magnetic resonance. No snooping tolerated. The bioengineering lab had an external connection to the Los Angeles Pediatrics Hospital and the Experimental Development Center, both separated from the lab by card-controlled access. Of the three key leading-edge programs housed within this complex, bioengineering was the star. The professor who had initiated the project had found and hired Jason, a recently graduated PhD in bioengineering.

At the age of thirty-eight he'd accepted a research position with the company. So he packed up his life, said a painful goodbye to his parents and close friends and got on a California-bound plane to Los Angeles. Only his excitement about his work prevented him from feeling that he had deserted everyone who had helped him. At the same time he knew how much he wished he had someone close to dispel the quiet and the emptiness he felt. He roused himself from his musings and decided to get some work done.

"Linda?" he called. "I'm going over to the hospital to look at those twins again. Okay?" No answer. His trusted lab manager must have left for the day. *Why aren't people more interested in this stuff,* he thought. *It's where our future lies!* What drove him in his research was the growing evidence of change in the DNA of the many donors who volunteered from the nearby college.

He left the lab, heading through the breezeway to the hospital to pick up results from the new volunteers who had also given DNA samples for the tests he was running. A subliminal thought, never quite reaching his conscious mind, slowed his pace. Was there a chance he might run into the dusky skinned doctor whom he had silently passed once or twice in the corridor?

Jason hurried down the passageway clutching two small vials. Twice he ignored the WET FLOOR notices and twice he slipped. Growing more excited—and oblivious to the risk—he picked up his pace. The thought that a broken vial would wipe out months of hard work lingered at the back of his mind, obscured by the memory of a beautiful face. The corridor turned abruptly left heading toward double doors hinged on the sides. In a fast trot now, fist clenched tightly around the vials, he hit the double-doors hard dead center, his shoulder parting them with such force that they clashed against both sides of the corridor. His shoulder hurt, but so what.

He didn't see the smallish person in a pink lab coat emerging from a door about ten feet farther down the passageway. Neither did the smallish person see him, but rather heard him coming and attempted to get out of the way. Ten feet wasn't enough. The pink-coated person clutched the folder of papers she'd been reading and froze, then quickly tried to back through the door she'd just left.

A look of resignation met a look of terror as the two grabbed at each other trying to avoid disaster. White coat, pink coat, black face, amber face, tangled arms and legs, shared pain, one dazed heap in the hall.

"Please let go of my breast."

"What?"

"Please let go of my breast!"

"Oh, sorry. Are you hurt? My God, the vials!"

Jason pulled one arm from under the pink coat and tried to sit up. "I'm so sorry. I was just thinking about someone . . . uh, something else." After a short but awkward pause, "I'm Jason, uh, Doctor Jason Towne from Genetics Research. Who are you?"

"Doctor Katherine—Kate— from Pediatrics." Her piercing blue eyes stared into his deep black eyes, dark pools of light, friendly light, she observed. "It's okay. You can remove your hand now."

"Oh, my lord, I really am so sorry!" Awkwardly pushing himself away from her, he sat up.

"What's in your hand?"

Slowly Jason opened his tightly closed fist not expecting to see the small vials intact. But there they were, completely whole. He let out the breath he didn't know he'd been holding. "Thank heavens they're okay. Are you?"

"Yes."

"Uh, nice to meet you. Sorry about the papers." Jason began to collect her scattered papers for her. "Ow! Ouch! Shish!" Rubbing his shoulder he awkwardly held out the fist clenched around the vials. "Nice to meet you. I've got to run." To himself, *Yes, nice.*"

If the rich deep black of his skin could have shown it, Kate might have noticed he was blushing. She stood for a minute, hand outstretched to empty air, then finished picking up her papers, still shaking her head. But she had a smile on her lips and a deep curiosity about what exactly was in the precious vial.

She dusted imaginary dirt from her child-friendly pink-and-white doctor's smock and tried to dredge up some memory of him from med school. Absently she checked the long rope of dusky blond hair tied neatly back from her face, hair that was congruent with her blue eyes but incongruent with her deep amber skin.

In the hospital corridor Jason checked the vials again. *Okay! Great!* He set them down carefully and continued rubbing his shoulder. *Kate,* he thought. *That's a nice name.* The image of her sunny appearance and calm manner played with his mind's review of the incident. She reflected the radiance of perfect health. A desert big cat of the Transvaal, he thought, lean and sensuous. He turned slowly hoping she wouldn't notice and caught his breath. She was still standing there fiddling with her hair. She turned, focused on the papers in her hands, and slowly passed out of his sight through the breezeway leaving him staring as she left.

The glow of the day's last light warmed the walls of Jason's office and lulled him to contemplate the fast flow of recent events. He'd considered making changes in his life before but hadn't acted on them until he had accepted the offer to pursue his research. It was guaranteed for long enough to determine whether his work bore fruit. He still felt guilty leaving his country and his family yet an opportunity like the chance to work here on his DNA research might happen only once in a person's career. His head hurt. Change had never been easy for him.

He picked up books and papers and headed back to his rented apartment near the genetics arm of the school. Halfway home Jason stopped dead in his tracks. The woman with the deep blue eyes drifted through some hidden portal in his soul and lingered about the edges of his conscience. He walked on not really seeing much of anything other than the faint image remaining softly in his mind. He entered his apartment and closed the door, dropped his papers and books onto the couch and headed slowly, very slowly, into the kitchenette. He wanted something cool. Jason opened the refrigerator and removed a soda. On the way back to the living room he suddenly stopped and scanned the room. A shiver passed over his skin, almost as if a warm hand had touched his shoulder. He looked again for the source. Nothing. Yet somewhere in his soul he had opened a door that had been long closed, and somewhere in his soul warmth began to grow. He sat down, leaned back into the comfort of an old armchair and closed his eyes. The woman with the cobalt blue eyes seemed to be standing in front of him. She turned her eyes to the stone amulet in her hand that her fingers explored. He'd seen that chain before, he thought, around her neck. It was silver, yet now she held it out to him, as it was broken. She was smiling. He looked into her eyes. Lights played within them sparkling in their depths. So deep was their blue that no pupil was visible.

A smile crept across his face. *Yes, yes indeed!* Oh, he had seen her before, the memory quickly transmitted as they passed in the corridor between their respective labs. Just a light sense of the fragrance of White Shoulders perfume, blue eyes, soft dark blond hair, warmth, and something he couldn't quite put his finger on. Maybe he could offer to fix the broken chain as a way to meet her again.

Kate had set about reordering the salvaged pages of her spilled notes. A doctor of pediatrics, she was aware that she could communicate with certain of her young patients, apparently through thought more than direct speech. She enjoyed her practice and didn't mind the extra burden of contributing her skills to the hospital-run clinic. Her innate gift of communication had become somewhat of a curiosity that she tried to keep low-key. Yet there were others who came to this clinic, and to her specifically, who also appeared to have the gift. She clutched the jumble of pages and headed to her car and home. The notebook she kept was private.

Kate flipped on the switch by the door as she entered. Music and light filled the room. Long days didn't dim her energy, and music and light lifted her spirits. For some reason she felt buoyant this evening. She dropped her sweater and briefcase on the couch, put the grocery bag beside them and executed a few graceful but energetic disco moves. Off came the shoes, off came the sweater as she flawlessly performed the rest of the dance while proceeding through the normal come-home chores. At the end of the piece, she went into her ample kitchen and began to prepare linguini with clam sauce and a green salad. Then she ate, still listening to music, still with a slight smile flirting with the corners of her mouth. None of this was abnormal for Kate but even she recognized it was somehow more than she usually felt after a long day doctoring.

Ten o'clock, time for a hot shower, a good book and sleep. She still felt as if a warm hand held hers. Slower now and sleepy, but warm inside, not something all that common after a long day split between clinic work and hospital work. She stood before the mirror on the bathroom door as she removed the rest of her clothes. Just checking, she thought. Twenty-seven was not exactly old maid territory. Piece by piece came off, and each was tossed in the clothesbasket or properly hung in the closet. She noticed the feel of her skin as her shift brushed across her breasts. She ran her hand lightly over the warmth and held her soft breast. Surprise replaced her smile, then a small thrill replaced her surprise. She stepped into the warm shower knowing well who had left the handprint on her soul.

Her mind drifted to an image of her rich bronze skin enfolded into his deep radiant black. The eyes of her soul melded into his, portals into eternity and connection with something new and exciting. She did not know the depth and extent of that connection and would not know until his research would one day reveal the secrets of the change. As sleep began to overtake her senses, quiet rhythms of ancient drums seeped sensually into her flesh, into her bones and into the dusky recesses of her soul. There are things that cannot be spoken, things too important or too sacred. His touch had struck a chord some place within her soul.

The next day dawned one of those soft, sensual, gray July days in California. Jason had been sitting in his office just thinking, wistful, longing for his own place. For whatever reason today felt like it would be a good day to drive around and see what California had to offer. His friend and colleague Greg had suggested the west side of Los Angeles where he lived with his wife Sara and son. Jason wondered wistfully whether Kate would like the same kind of house that he would. *There I go again! I don't even know the lady, why am I thinking like this?* Jason admonished himself to be rational. Yet, in the back of his mind Kate would not go away. That was okay. It was comforting to him to have her remain on his mind.

When he had first come to the area after accepting the job at North Western Jason had hired a cab to drive him around. He needed a little introduction to places of interest including a side trip to places he might like to live. They had ended up in the Santa Monica Mountains where the mountains tapered off toward the Pacific Ocean.

Today he would have another look in the canyons of those mountains. One area in particular reminded him of the more open country of his homeland. It didn't take long for Jason to find himself driving toward the hills, nor did it take long for him to locate the canyons of his interest. This area had several fan-shaped canyons sweeping toward the Pacific. There were three main runs to the sea, each softly sloping, large level patches, all less rugged than those farther inland that had captured his eye. Two of the three already sported houses, one of which belonged to Greg and his family.

Jason's gaze wandered off into the distance and came to rest on the ocean. His thoughts drifted to the softness of the day and the cool marine layer that had crept in during the night bringing relief from the summer's oppressive heat. A California red tail hawk in flight caught his eye. Two others, larger than the first, joined it. The smallest of the three maintained a screeching diatribe aimed at the larger two who intermittently appeared to assuage its complaining. He laughed. To him this day was one of those silky days that can either inspire one to a game of golf or leave one sitting in a lawn chair pretending to read a book. He wondered if Kate would like the location as much as he did. He decided to invite her to lunch and see if she would like to take a drive out there.

Jason strode into his office a few days later grabbing the contents of his in-basket as he passed. A travel brochure from the pile crumpled in his hand as he groped for his five-legged chair. More than once that *thing*, foisted on him by the caring ergonomics folks, had entangled him in its five splayed legs that fanned out from a wobbly stem supporting the seat. Its two elbow rests, which he could not bring himself to call *armrests*, were never where he thought they were. "Linda, I want a new chair!" he bellowed. "One I select!"

"Okay! Okay!" Linda sucked in her breath through clenched teeth and reflexively reached out with her free hand trying to avert his impending disaster. To herself, *Lordy! Get him out of here!* To Jason, "Here! Take the brochure. Take a trip to Australia. It'll be nice in the fall!" Why a five-legged office chair confounded her boss escaped her. It was carefully centered on a carpet-saver pad and with plenty of room for a normal person to seat himself without a major confrontation. In all the time she had worked with him she had never been able to avoid a sharp little gasp as she watched his seating ritual with *the thing*. And never once had he failed to glare at her for her watchfulness, a confirmation of his own inability to focus on more than one thing at a time.

Once stabilized Jason continued to test the chair. He eased himself into it, managing a small apologetic smile. Linda moved toward the door. "We can go over the last print runs tomorrow," and she was gone.

Jason punched the speaker button on his phone and pressed *play*, briefly listening to messages. Nothing urgent. The slick paper of the now crumpled travel brochure felt good in his hands. The ink on the paper had been scented slightly, smelling of the outdoors. *Good trick*, he thought. *Australia, the great virgin continent that draws people to her. What a place for a honeymoon!*

Jason grabbed the handset and dialed Kate. The phone rang once, then twice. She must be busy or out, he mused. He put the brochure down on the green desk protector provided by someone who respected the antique mahogany desk as much as he did. The chair he hated and in no way at all did it belong with his family heirloom desk. He loved that desk. It had been his father's, and before that his grandfather's. He'd had to fight for its right to be in his office. The dark carved wood of it, its massiveness, all the drawers perfectly balanced, and even a couple of secret drawers, the smell of the wood, all put him at ease and defined an inviolable space that was his. He could work here. It was at this desk that insights into the intricacies of genetics associated with the paranormal had occurred.

He tried the number again and thought of spring in Los Angeles being autumn in Australia. Then he heard Kate's warm, "Yes? This is Kate."

"How about two weeks in Australia?"

"Sounds perfect! Let's talk tonight over dinner."

Days spent together melded into weeks during which they took long walks, enjoyed leisurely dinners and basked in the warm summer evenings together. The hills and canyons of the Santa Monica Mountains interested Kate almost as much as they had Jason. The fan-shaped group of three canyons had two large lots available. The middle canyon belonged to Greg's family. Jason and Kate liked the two on either side of Greg's. One had never been built on, but the other house belonged to an older couple that wanted to move. They walked the canyon slowly and discussed the virtues and drawbacks of starting with an existing house versus the hassle of constructing one. They were both excited about fixing up the older folks' place and not so thrilled about new construction. Besides, they could live in the house as they converted it to their needs.

In their excited plans for the future they suddenly realized that neither had proposed the topic of marriage. They stopped laughing long enough to walk down a path to a small gazebo overlooking the sweep of the canyon toward the ocean. The decision on the house made, they took turns proposing, Jason first, then Kate—when she stopped laughing.

"Australia!" they blurted out simultaneously and moved into a spontaneous long-lasting hug.

"Jason," Kate said as she gracefully extricated herself from his bear hug, "I wonder if your friend Greg would be home? We need to introduce ourselves as their future neighbors."

IV — SEEDS

And from their ethereal perch in the heavens, two Light Beings viewed the couple.

"The journey begins tonight, OmRa, with the first of the four chosen tablet bearers."

"Yes, Thiesha. As evening moves across the Earth, the other dances also begin. The music of the ages accompanies them all."

The lavender seed of light hovered softy, unnoticed by the man and woman listening quietly to the soft, slowly pulsing rhythms of the didgeridoos. Together, they were unaware of the lavender cocoon beginning to spin around them, but the Guardians knew the new life to be formed here would be special. Slowly, the intricate beat of polished sticks tapping together began speaking their ancient messages in rhythm with the throb of the didgeridoos. The group gathered in this ancient unchanged land, witness to the secrets of evolution, awaited in silence for its part in the dance of creation.

As dusk became night, the lavender light pod continued forming itself around the couple now lost in the perfume of each other's body, oblivious to their surroundings. Mahrianne stood by, guiding the play, each new performance differing only in its participants and locale, always exciting, always one step closer to completion of the play begun so many centuries ago. She parted her lips and lifted her arm, palm forward, signaling the lavender mist to rise, carrying the enjoined pair with it. As she raised her hand over her head, the soft folds of her garment slid down to her shoulder baring her bronze flesh, its color enhanced by the glow of the fire. A quick turn of her hand and the lavender pod began its slow spin around its vertical axis.

As the pod of spinning light turned, on-lookers could discern a bright silver-blue seed take form near the pod's apex. The silver-blue seed of light was pulled to the center of the light pod, the couple therein no longer visible to the watchers. As the pod picked up speed, a haunting chant, primitive, unearthly in its sound rose from the group, quietly at first, then rising as the pod sped faster and faster. Didgeridoos joined other ancient sounds to blend into a resonant blanket of celebration, expressing knowledge that this dance of creation signaled the near completion of a task begun two thousand years before. Suddenly iridescent greens and reds swirled within the lavender pod.

Jason became lost in the warm forest between his mate's soft thighs, barely noticing the bluish mist that enshrouded them both in a light fragrance of peace and fulfillment. Something new, something magic was happening. As he entered her welcoming flesh, a flood of beauty and warmth swept over him. The joy of her, as always, came naturally, wrapped in soft sounds and enfolded in the warmth of her arms. Within the woman the silver seed selected one of the seeds of the man and moved it swiftly to its destination. As their contributed strands of DNA unwound, two identical strands of a new-format DNA entered the genome, and as these unwound, the steps of the dance of creation took over. The ladders of the unwound strands moved between the ladders of the male and female DNA.

As if by command from somewhere, Jason opened his eyes and looked into hers. He felt rather than saw her blue eyes become the black of the eyes of all space and time. At first her eyes were open, surprising him. Their bodies moving as one, he did not know what had made him open his eyes. At this moment their connection was total and profound. What he saw was the shape of her eyes, slowly, almost imperceptibly elongate and close to slits. Slits into which he fell bit by bit, into the black abyss of space and time, bright spots of light, stars, galaxies sped past as he sank deeper, deeper into the space that was her soul. The dance of their flesh joined the swirl of the abyss, warm, glowing, faster, faster, somewhere, no fear, only awe and excitement. They spun together, consciousness joined in the knowing of oneness. Atoms indistinguishable by owner, joined in an experience beyond the veil of life, a creation only partly known, willing, ready, a new knowing that all space and all time are one, and infinite.

Her blue eyes now black slits of infinite depth and light, doorways to infinite universes within universes, peaks and troughs in the standing waves of time, all sustained through the great *om* of space-time, echoed and invented, vibrating, reflecting, joining in the mysterious space of the gods.

Successful, Mahrianne continued on her journey. She arrived once again, quietly, a wisp of cool air, the scent of woman bringing comfort and safety, readying the place for conception of one of the special four, one of the tablet bearers.

She smelled of corn silk and blue lupine, fresh from her bath in the wide sandy shallows of the great river. He watched as she dried herself in the warm sun, oblivious to his presence. Not that she would mind. They had been joined together now for many moons and her heart sang as she felt his presence.

Pretending not to notice, she moved quietly to her own rhythm spreading her shining black tresses as she moved slowly, her dusky red skin vibrant in the Sun's rays, calling him closer.

Moving to some inner rhythm, pulsing, sensual, wild, her soul screamed, twisting and crying—for what? A voiceless whisper, her hair sat restlessly on her head, twitching a little, she thought, causing her to push at the roots to make them be still.

He could see her hair from the top of the bluff, blowing, long, black, silken, wild. Sunshine coalesced haloing her features, almost indiscernible from this distance. Echoing, sunlight glinting off of her shining hair created swirls of bright corn-silk yellow, a reflection as she moved her head.

Overwhelmed by the magic of her, something he had never anticipated in their union, Gray Wolf imagined the soft warmth of her body pressed against his, sinking more deeply into some spell he never quite understood, but never resisted. At times like these, their bodies made to the measure of the other, he succumbed to this feast of delight, feeling, smelling, tasting, entering—timeless, and he marveled at the plenty of their love.

Not the many times they had joined together, not the quiet warmth of each other in the night, not all of the magic of the many full moons they had shared had dampened his great thirst for her. It was not the matching of their closeness, it was a joining of powers beyond their simple understanding that drew them together.

The Old Ones watched. Faint shadows of the smiles of ancient wisdom played behind their still lips. They watched, unseen, for they sat joined in a circle atop Black Mesa, seeing the joining that they knew would produce the tablet carrier, the messenger of the reunion of man, the long, long awaited gathering of the tablets.

Sun caught each droplet of water sliding from her skin, each diamond of rainbow color she flung into the air as she dried her hair. Gray Wolf lay down his deerskin quiver of arrows. Quietly, tilting his head slightly back he mimicked the call of a flicker, a desert bird seeking its mate. She did not turn for she wished to hide the broad smile spreading through her soul, warming her waiting body. Swiftly he made his way down the bluff. Moon Shadow waited, feeling rather than hearing his approach until she could be sure he was there. As he reached out to touch her shining hair, she turned, its long strands following her turn, settling sensually over her shoulders and reaching to her waist, only her erect nipples visible, inviting, beckoning him forward.

Their embrace rang through still open air, silent, magic, blending. On the mesa top, eyes closed, the Old Ones joined hands sitting cross-legged in a circle, moving silently to one side, then to the other, listening to the voices of the ancient ones, those who had passed on saying *This is the one, he has been invited.* Their loving song blended with the song of the infinite. Their eyes became the eyes of the universe, and in the background the slow steady beat of the soft drums overlaid with the chant of the Old Ones heralded a new dawn for humankind, a new beginning for precious life.

Mahrianne stood with the Grandfather, watchful and protecting, for nothing must interfere with this union.

The lovers lay still, man and wife, entwined, sharing their secret awareness of the strange and beautiful act of conception, each cognizant of something rare and meaningful that was transpiring there below the mesa. The bond between them was a bond of forces far greater than the two, a trust for all of humankind. They were unaware of the soft lavender light swirling around them or the changes being wrought on their progeny. Neither were they aware of the presence of Mahrianne guarding the two and the changes being made.

Grandfather-to-be rose slowly, closing the circle by meeting the hands he clasped together, and walked slowly to the edge of the mesa. There he opened his eyes and looked down at the young lovers oblivious to his watchful eye. No harm would come to them as long as he or any of the other Old Ones were alive to protect them for Moon Shadow would bear the tablet carrier.

Gray Wolf pulled her closer, her head now resting in the curve of his neck, his slow steady breath warm in the forest of her hair. There she wandered, pleasantly lost in the thickets that led to sleep. Her eyes closed for a moment while the seed of new life found its path to its waiting chalice. Only barely did she notice a blue and lavender light around them.

Once in a dream, though she had no words with which to think about it, Moon Shadow had been surrounded by blue and lavender light. She knew her role, her special place in the ancient drama that would unfold. She had been touched by Mahrianne.

Gray Wolf lifted her hand as they watched the magnificence of the desert sunset. As it faded, hand in hand they arose and walked up the long trail to the surface of Black Mesa. They watched the radiant sunset then stepped into their pueblo as they had done so many times before. And as they had done so many times before, they lay side by side, their arms reaching out to each other for warmth and comfort. Her bright black eyes searched the waning light for his, and as they locked into each other, a thrill of knowing swept over them. This was their time. They had received their gift. The evening brightened imperceptibly with a shimmering blue light containing all the stars of the universe in their infinite depth.

The old wise man walked with measured pace along the edge of the mesa, Grandfather of the tablet bearer yet to be, sun-lined face fixed on something unseen in the distance, shoulders stooped with age and the burdens carried for humankind by his people. He paused facing the sacred mount to the west now crowned by a spirit cloud. As he passed the house of the couple, he made a gesture of blessing for the pair.

Near the edge of the mesa, he sat down, cross-legged with a grace belying his many years. He sat thus, head raised to the heavens, and gently held the wind flute. He caressed it softly as if to blend with it, the caresses worshiping as if he were caressing the long-awaited newborn child, the child bearing a special gift from the Ancient Ones, a gift for all humankind.

No sound could be heard. Quietly he drew a breath, a deep sigh from the wind, and once again placed his fingers fluidly over the openings of the flute to join a ceremony held in another realm somewhere in the stream of the infinite. The mesa was silent save for this. The notes floating from the instrument told the story of fulfillment of promises for a new human race. Stories of the Hopi Nation had been passed down by word of mouth generation after generation, telling of the destruction of the first Four Worlds and the opportunity of one last Fifth World—this to be the sacred trust of the Oraibi Fire clan. This last world would be enjoined when the four tablets were finally brought together to tell the role of each race in the Life Plan.

The haunting notes of the wind flutes floated out across Black Mesa, dripping down the edges of the mesa onto the desert floor. Soft pulsing rhythms joined as muted drums compelled bodies to move quietly in rhythm with the ancient music, music bearing sounds of longing and waiting, slight qualms of anticipation of some event passed down generation to generation. These sounds carried the message of deliverance and duty, a covenant understood, the preservation of the old ways through trust, knowledge and stories yet to be fulfilled—the only means to the survival of humankind. Thus was the solemn duty given to the Hopi to live one way through all time and to preserve what the ancient gods had given the Hopi as a sacred trust. Now the tireless trek through life after life, soul after soul was ending. The faithful spirit guides, owl, hawk, eagle, wolf, bear and buffalo were always there, always guiding, always encouraging, keeping faith through the centuries, carrying yet a great sadness with their joy and sacred trust for humankind.

The Earth continued on its path through the heavens, turning as it floated, slowly bathing the other side of the world in sunlight. Above the mountainous realms of Asia and Europe the Light Beings again watched the swirling lavender mist and heard the lilting music of the ages. The dances continued anew, promising fulfillment of the legend and giving hope for the future.

V — THE NEW ONES

"They have chosen to return to the place of conception to celebrate the birth," mused OmRa.
"It was preordained. They do not need to know why."
"Their progeny will represent the black race."

Jason's daughter came out of the chute ready to rock-n-roll. At 5:01 a.m. precisely her glistening chocolate-colored head popped out looking very much like a wet seal pup. The rest of her perfect baby body slid easily into the waiting hands of the aged mid-wife who grasped her firmly, one hand clasping the tiny feet, the other supporting her back and head. The birthing sounds created by the symphony of didgeridoos, drums and accompanying instruments eased from the pulsing rhythmic chant of deliverance to a soft, soothing lullaby. Another woman counted slowly in musical tongue then swiftly severed and tied the umbilical chord and began coaxing the afterbirth from the mother.

Even so, the impatient newcomer took charge. She did not wail, she bellowed. Just a small bellow, but an unmistakable, *I'm here! Who are you?*

The big grin on the mid-wife's face displayed neat white teeth despite her advanced age. She laughed out loud, slightly bouncing the girl in her hands, welcoming her to a new world.

As the first bright rays of light spread across the sky, the child's eyes opened even wider than before, trying to focus on this new place. Jason was handed his daughter whom he accepted with tears, eyes bright with pride. His smile quivered ever so slightly as his love for this child overwhelmed him. Strong man-hands helped him to a resting place where he could sit with his daughter. They handed him a vessel of fragrant oil with which to clean and protect her delicate skin. He spoke to her softly welcoming her and reassuring her that she was safe. He thought the cooing sounds she made were her talking to him. Maybe they were.

OmRa chanted quietly, a singsong chant bestowing gifts upon the newly born. "Born in the first Moon of Earth renewal, the snow goose shall be her spirit guide. Indian Spirit Walkers have sent her the great red-tailed hawk to tap the energies of Earth and guide her through their spirit."

Thiesha joined the chant. "Her reserved, serene manner shall serve her well as a receiver and transmitter of the great powers of the universe. The heavens are her realm; her mind connects with distant places."

Her eyes were deep blue, of a hue so deep as to be almost black—pools of eternity wherein blue lights played, now reaching, rushing out as a flash of blue, if an observer were quick enough to catch it, then swirling into deep pools. Deep pools in a face of ethereal beauty, gateways to other realms— open, calling, pulling. Eyes lined by thick lashes, cilia alive as reeds about a pond, cool, thick, moving gently in some unseen wind.

Jason almost dropped the child but put her down, reluctant to peer into those dark pools anymore but also to avoid being drawn closer to her until he could comprehend the magnitude of what he felt. He knew she was different, even more than he had expected from his research and pondering on her genetic material, but how much was yet to become apparent.

Her tiny mouth curled slightly, quivering at the corners in disappointment. His face reflected his own feelings in deep contact with this small beautiful creature of his doing and now about to cry. His heart became one with hers as tears muddled his vision. Her sorrow vanished as strong gentle hands again lifted her up to his face and his warm soft kiss graced her forehead. The curl of insecurity transformed into a forgiving smile, safe in knowing her connection was made and he was hers. She was anxious for more of this place and the safety of this great creature's hands.

Soon the mid-wife came to him, gently removed the babe from his large hands and carried her to her mother. Kate was propped up now comfortably leaning on cloth-covered soft grasses, clean now and free of pain. The mid-wife placed the baby face down on the bare skin of Kate's chest, her tiny face resting against one of the swollen breasts. Kate's hands covered her child and the sound of her voice poured over the infant, covering her with love and welcome. Her voice resonated within her chest and soothed the child who, now wide-eyed but quiet, began to grasp the warm flesh of her mother, kneading it in her tiny fists.

The chant of the didgeridoos now softened. Accompanied by ancient prayers and songs offered by the gathered, the beat gently slowed. Slowly, softly all came into rhythm with the new mother's calm heartbeat.

Jason, now beside his wife and child, closed his eyes. He placed one arm under Kate's head and shoulders, the other at his side with his hand covering his daughter's body, cradling them, sheltering them with his deep love.

And the mother and child slept, watched over by Mahrianne and the quiet keepers hovering mistily in the dawn light.

Rested at last, Kate stood, holding her infant. The smile on her face reflected the deep pride in her heart. Jessica had arrived. For Kate, it did not seem odd for an infant to have such a commanding presence. But indeed, Jessica clearly held center stage and thoroughly enjoyed it. A wide grin played upon her toothless mouth, sweet gurgling sounds coming from it as she assessed the staring faces, trying to focus her bottomless eyes. She felt their love and awe as a passing thought, *I am the master here*, and she puckered up, huge tears welling in her big bright eyes as she announced her hunger. *Look at them! I am the master here!!*

In a mountainous kingdom at the top of the world, the second tablet bearer was awaited.

"We now have one race represented, Thiesha. Let us attend the next, that of Earth's yellow race."

"We have traveled far, OmRa. Let us see what Mahrianne has brought us from this mountain kingdom."

For some time the deep resonant voices of the holy Tibetan monks had been chanting the evening devotion for the collection of followers who had joined the procession. Group by group, small numbers of people had joined with the caravan that provided some measure of security for the many people seeking a better life far to the west.

As the caravan drew slowly nearer the ruined temple, a young woman moved farther toward the lead group of the long procession. From the beginning of the new life growing within her, some ancient wisdom had marked the child as exceptional. People brought gifts and prayers for the child and beautiful scarves for her. Her skin echoed some ancient call and she hastened on the trail ever upward, passing curved dwellings wrested from the stone itself and then on to the ruined temple. Trees grew there now in the remains of the beautiful temple gardens. A static charge passed over her sensitive skin. She looked at her arms noticing the few pale hairs moving when she passed her hand over them as if some magic now surrounded her.

The caravan slowed and stopped, ready for the evening's rest and preparation of the evening meal. With the two women who made the journey with her, she moved slowly into the garden and the friendly old trees of the temple. They made ready a small birthing area in which they would greet the new child.

The wind soughed softly in the trees and brushed gently across her damp forehead. Longing for the gentle touch of a lover, she calmed as the women soothed her hot body. Her mind began to drift in the caressing of the breeze, lifting slowly from the tender young grasses in which she lay waiting for the rhythmic movement of her womb. Gently, now firmly, shifting, pushing, coaxing the new life from its safe haven in her belly, she gave to her body and the women the task it was designed to do for she was new to its realm.

Her arms stretched upward, her mind catching the top-most boughs of the trees. Her skin, so exquisitely sensitive to the slightest touch, thrilled to the gentle spattered caresses of the leaves. As she moved slowly to face the fading sun, a small movement caught her eye. Her mind moved out to touch the small thing coyly perched upon a nearby branch, watching. She thought it was a child. The apparition there spoke to the mother in a language without words—*I am your son, I have come to help. We have time. We have much to do.*

The thread of his mind wove itself into hers, and in her mind she felt a swift downward passage through the rustling trees. It ended with her kneeling on the soft grass, legs as far apart as she could make them, the beautiful head of her newborn son passing swiftly in one joyous rush. Her now bloody hands held the boy, head cradled gently in her left hand, tiny buttocks in her right. They cleaned him with the tender grasses as she wept tears of joy and creation, unable to contain herself and bursting with pride at his perfection.

The child did not cry but cooed softly, appearing satisfied with circumstances, wiggling in his newly found freedom. Born in the first Moon of Earth renewal, the snow goose would be his spirit guide. He would be the messenger conduit between the spirit and non-spirit world. As a visionary, his dreams would be prophetic. Using these he would be seen as a humanitarian.

When the afterbirth passed, the women briefly inspected it then scooped enough earth away to hold it. Then they quietly thanked it for the work it had done and replaced the grasses they had moved to cover it.

She laid the babe on her chest, leaned against the warm trunk of a tree, and they slept.

In an alpine land far from the other mountainous kingdom, the third tablet bearer was awaited.

The young woman stood in the kitchen, a paring knife and a potato in her hands. The first pangs struck strong but not sharp. The paring knife she had been using fell from her hand into the sink. The potato slid into the opening of the sink and lodged there. She did not notice. Her clear blue eyes momentarily flitted about the room searching for the cause of the twangs, then she laughed at herself for forgetting her time was so near. Sweat began to bead on her forehead. Nothing out of the ordinary, it was her time. She wanted to be on the mountain at her special place, a quiet protected spot that looked out over houses and streams and on into that special mist that sometimes graced the valley below.

She tugged on her mountain boots, laced them tight and picked up the backpack she had prepared for this moment. Then she headed for the trail that wound its way from the valley floor up through the farms, the herds of sheep and on into the stately alpine mountains.

The forest creatures made no sound as she passed. No breeze rustled the branches of the silent forest. Waiting. Waiting. She paused to lean for a moment against the trunk of a great tree shaded by its spreading branches, cushioned by grass and lichen. Waiting.

Somewhere a small songbird whispered several cheerful notes, bringing a smile to her dry lips. She lifted her water jar and drank from its cool content, then moved on up the mountain toward her sanctuary. Again the wave of birth began moving from the top of her full belly, coursing in waves down through her body and ending low with a tugging compelling stretch. Her breath came fast, in short pants, fighting the pain. No full lung of air could be had. As the wave passed, she felt relief and wiped the moisture from her brow that had gathered there. Her time was close and she wanted her private place. She forced herself to move on, each step closer, each step heavier, and each step more taxing.

Finally she was there. She sat on the soft grasses, unlaced her boots and pulled them off. Then she lay back breathing hard and pulled the fragrant air smelling of June flowers and green trees into her hungry lungs. A soft breeze arose from somewhere down the hill, crossing the chill mountain stream at

37

her feet and soothing her hot face and body. She stretched out her legs and let her feet bathe in the cool of the stream.

She sat like this for some time, each wave of her birthing more tiring and stronger than the last. She knew whatever miracle occurring now was about to unfold for her, something she had watched before, but this was hers. The lavender haze forming around her was somehow familiar, but not so she could grasp its significance. She felt but could not see another female figure. A gentle familiar wave of sounds played about her as, unnoticed, Mahrianne performed her role.

One more swallow of the cool water in the jar. She set the jar down turning at the sound of footsteps running up the path to this spot, her private spot to which she, all of her life, had retreated to be alone. The runner was breathing hard. It was a fair distance from town up the mountain to this place. Almost no one ever came there. He appeared at the break in the trees, breathing and panting. She chuckled. He looked pathetic, but worried enough that she felt guilty at not telling him her time was here.

He knew she was frightened of hospitals and had guessed the spot to which she had disappeared. As he reached her he regained some of his breath. He hugged her gently and managed a soft whisper in her ear. "I am here. May I stay? She is mine, too. Please? You are mine, I can help."

She lifted her arm and touched his face. Only a small smile could she find for him. She was otherwise occupied but whispered it was time. He gently lifted her away from the supporting tree and held her kneeling, facing the cool breeze to await the arrival of their daughter. He knew no argument would get her down the hill and to the sterile white of the hospital.

She raised her face to the ancient tree feeling its power take some of the weight of her body. She felt the softness of the cool grass and lichen about her legs and surrendered her spirit to the forest and to the towering mountains rising from the Earth. The white snow of the mountaintops stood vigil over the quiet landscape. Not a sound anywhere. Her breath caught in anticipation. The warm wind coming up from the silent valley lifted her spirit and with one freeing move the girl child entered the world.

Born on the first Moon of the strong Sun, the deer would be her spirit guide. Indian Spirit Walkers also sent her the wild rose to tap the energies of the Earth and guide her through their spirit. She would be gifted with mystical talents and an all-healing love, she would be reliable, analyzing before acting. Peace and stability would be her methods.

He held the baby in his hands, tears of joy misting his eyes, then laid the girl gently on her mother's belly, now rid of its charge and soft. Using a twisted length of strong dry grass he tied the cord, cut it and waited for nature to finish her job. Mahrianne sang softly with the breeze, spinning an aura of safety over the three.

The baby's sudden cry rang out as her small fresh lungs filled with the sharp air from which she would now draw life-sustaining oxygen. The sound surprised her and she bellowed on, engendering laughter from her parents as well as tender sympathy and instant cuddling. At the melodic sound of her mother's soothing words, she tried to focus eyes unaccustomed to use on the source of the sound. Failing that, she succumbed to a yawn and sleep. She was tired. She had had a big day.

They sat like this a long while. The young mother lay holding their daughter cradled in her arms, exhausted and content, glad he had come, and he watching to be sure they were all right.

The wind blew the spirit name, Mahrianne, to those waiting on the mesa, a call to guide the long-awaited tablet bearer's birth.

"The final birth awaits. Wise hands will guide him on his path."

"The spirit world is strong here. I have great hope for this tiny world."

The silent breath of the great owl's wings brushed Moon Shadow's hot face. She was startled at the creature's huge size as it stopped abruptly, landing on a branch of a scraggly tree directly in front of her. He uttered no sound as he adjusted his majestic feathers and settled in on the branch, eyes unblinking, focused directly on hers. It seemed to her that he was the messenger bringing her notice her time was here. She smiled at the great bird, it softly replied, a throaty muted melodic sound.

The whole clan had become more and more available to Moon Shadow in the past week. Little things she usually did somehow got done without her. Her joy at the honor of being the bringer of the tablet bearer had long since switched to weariness and anxiety. Now she became focused on one thing. She and her babe were special, and her anxiety turned into total trust in events to come. She shared the clan's pride of the importance of this child and she understood that security around his existence would be necessary. Only the Hopi shared both that knowledge and the responsibility. There would be no failure.

Hollowed logs, hard wood, and drumheads of animal hide, all beat rhythms of ancient times. The flute players, trance-like, breathed a stream separate from the sounds created as breath passed from them through their instruments. Over it all, spirits wafted, guiding, watchful, and everywhere present. They had been called by the flutes, brightly adorned with feathers and paced by the intricate patterns of hard sticks on hollowed wood drums of varying timbre and depth.

Mahrianne moved to the sounds growing with a life of their own, a living formless organism, captivating mind and body, joined by other drums and wood flutes on Black Mesa. One clan, then another, joined in, all playing the same ancient sounds. As they heard their neighbors over the local sounds, the other clans entered the ceremony and joined in the chants as they could hear them. It mattered not. The sounds produced all began the same, all contributed to a mounting edifice of body and mind and of souls yearning for rejoining. Piercing in its simplicity, the breath of the clans merged, rhythm in, rhythm out, all driven by the accelerating pace of the sticks, drums and flutes.

In Oraibi, first one then a second and then another joined the birth circle. Moon Shadow lay on a raised mat of soft skins and herbs, fragrant and aromatic, facing the players as they assembled. One after the other, the women danced slowly behind her, lending to her their great strength and gentleness. A tiny silver bell bearing ancient carvings and passed down through time had been saved for this purpose only. The women took turns tapping it with a small stone mallet. The silvery tones issuing forth told the story of the one who would play a strong role in the fate of humankind. For this was one of the special births, and this one, very special.

Behind the men and their instruments, the children walked slowly back and forth, now performing intricate patterns, now weaving in and out in snakelike ribbons. Eyes closed, their arms and hands made smooth weaving patterns as their bodies bent and swayed to the rhythms. Ointments were smoothed on the mound of the new mother's belly to ease the movement now overwhelming her.

Mahrianne, keeper of the new one, watched from the side. The red came in the night—tears for humankind, joy for the world. The birth had announced its presence. Mahrianne whispered to the gathered, "Still your hearts and listen. He comes."

In the distance eerie notes, echoes from other times, haunting, penetrating, ancient, rose and then drifted into the dust of the desert below. No other sound was heard save the slow haunting tones of the ancient flute spirits passing encouragement through the old Indian's soul and through the wooden wind flute. His fingers moved softly, commanded by the Ancient Ones as their voice, for the children of Earth were far removed from the path, yet they clung to things they did not understand yet knew their truth. They had accepted the heavy burden to preserve humankind's last chance. They knew the sense of urgency. They knew this was the time they would be called upon to execute sacred rites handed down through millennia, a hand extending to men of today to take or to perish. All wept silent hot tears of acceptance, knowing that now was the chosen time. The honor was theirs.

Soft rain began, the wind a soft sigh rushing before. The old flute player, Grandfather-to-be of the child being birthed that night, shivered with a thrill of excitement. He knew not why. A low moan from the new mother told the waiting that the child would be brought in by the light of the dawn. In the miracle of the dawn, Venus, still bright, gave way to sunrise. The heavens celebrated spreading the sky with clouds of color, peach, apricot, and then touches of lavender.

A young red-tailed hawk circled the gathered, deftly playing on a growing thermal, waiting for the sky to turn flaxen. The plaintiff cry of the hawk broke the soft hot air as she circled the mesa; her bright red and rusty brown feathers gleamed brilliantly against the bright blue of the sky. Her lonely cry enjoined those waiting on the mesa, *Where are you? Where is my mate? Warm breezes wait. Come fly with me, come join with me again.*

On the mesa no one moved. No one made a sound as the eyes of the hawk and the old man met, eyes fixed on and in each other. Softly a chant began, a murmur from the souls of the Old Ones gathered round. The sound of times ancient, almost forgotten, echoes to tales passed down through generations of waiting, passed on through one link to the next, all melded into one urgent request. *Let this child come, we have no more time.*

The hawk playing on an uneven thermal waited for the Sun to flash from the soft colors of a pending dawn to the brilliance of the explosion of dawn, bright with a light of its own making. As the dawning sky flashed flaxen in synchronism with the burst of the Sun, she screeched a long exuberant cry, joined by the mother in her final push. The old woman attending her wrapped and held the babe. In the miracle of the dawn, the male child breathed.

The old Grandfather's heart ached, almost imploding with the pain and burden of humankind.

Sheet lightning flashed silently against the dawn. Carefully the old woman held the child while the withered hands of those wise elder women of the clan cleaned and soothed the child.

Mahrianne remained to the side, watching, protecting.

"*Born in the first Moon of the budding trees,*" chanted OmRa, "*he shall be purveyor of spiritual evolution. Intensely intelligent, the fire of the life force shall burn brightly for this friendly, gregarious boy.*"

"*The sky is his realm,*" enjoined Thiesha. "*He shall commune with the Creator Spirit and cleanse the ideas of those he touches.*"

Gray Wolf smiled at Moon Shadow. Gently he lifted their son to face the new dawn. "Wind Runner!" was all he said. It was the most he could manage lest he show too many of his deep and profound feelings. Her nod was sufficient. He raised the boy higher and spoke the name to all present. Silently, to only himself and his mate, he gave a strange and prophetic prayer that their son would lead humankind to survival and the gift of a world in which to flourish.

VI – Children of Transition

The swirls and twists of time and space reflected in the babies' eyes. As with other children these days, some difference showed itself through the unusual and unique patterns of colors. Or was it reflected light? Reflected light would appear as random patterns or flecks, but always similar. " . . . *it's just the way babies' eyes are,*" people thought.

But that was not the way it was. Eyes that appeared usual and alive on one day would, under different circumstances, alter in some unexpected way, ways that led to infinite complexity should the observer look more closely and amplify the patterns in the eye. To do so would have shown strange and beautiful patterns of multicolored paths infinitely repeating, one spawning another, identical, growing and repeating over and over— fractals, equations generating patterns that stopped and seeped into and from other dimensions, maybe even other universes.

No one completely understood these windows into the souls of the children now being born. They more likely thought of them as complex cogs in the wheels of change. Perhaps they were waiting messengers and conduits to a higher wisdom. Or perhaps they offered assistance, planned and proffered to save a humanity unwilling and unable to love itself enough to take the time and measures to address the inescapable knowledge of the great change fomenting within Mother Earth? This was the danger of just avoiding dying rather than living. Or perhaps they felt caught in a cataclysm so widespread and intense as to be incomprehensible, thus denying the destiny of Earth, avoiding preparation and education to assist in its hours of need. Who will remain? Who will be left to assist? Volunteers?

"Congratulations, Jason!" Greg patted his friend on the shoulder as they walked towards the lab. "I hear baby Jessica is doing well."

Jason's hesitancy caught Greg off guard. "Why so thoughtful?"

Jason stopped and faced him. "We've been friends for a long time, right?"

"Of course. What is it?"

"Well, this is going to sound really strange." Embarrassed he looked away. "Come on, let's walk." They continued on down the hall. "You know that Kate and I were both drawn to Australia for our honeymoon. I suspect Kate knew somehow that something extraordinary would happen. But something was different about the conception of our daughter." He paused. "I told you this was going to sound strange, but is it possible that there was an intervention of which we were unaware?"

Greg cleared his throat quietly. If Jason weren't such a close friend, he would hesitate to ask about personal things, but in this case . . . "Tell me exactly what you mean, Jason. You know my background in parapsychology. Start at the beginning. I'll make the connection to other realities if I can."

Jason began quietly, allowing himself to transfer his troubling thoughts into words. "It was an incredible event in our lives, nothing as we expected . . . "

"He knows, Thiesha. He does not yet want to acknowledge it, but he knows."

"It will manifest itself in due order. He will observe her growth."

At home that evening Jason watched the child, small hands grappling with small toes. What was it he had seen there in those unearthly eyes? What had been communicated from such a new mind into his? He knew she was his, his and Kate's, yet, things he had ignored before began to pop into his head.

Startled by an abrupt disconnection of her hand from her toes, baby Jessica jumped and threw her arms out to her sides to brace for any possible consequences of the sudden disruption of her pleasure. Small quivers played around the corners of her mouth and the blue-black pools ran over, trickling down the sides of her worried face.

Jason, overwhelmed by the tragedy just occurred, quickly scooped her up, holding her upright against his shoulder. He patted her gently on her back. Her fear became his sorrow and he realized his own hot tears had filled his eyes. Her feelings had transferred to him! They remained thus for some time unaware Kate stood in the doorway, a soft smile on her lips, watching the big man holding his small tawny daughter with the infinite eyes. Little did he know, she thought, just how connected they were and would always be. As the man turned and began to pace with the child, his back presented a small teary face perched on his shoulder to the woman in the doorway. She smiled broadly at the child and winked at her wee daughter. The little head lifted a bit and smiled back, but quickly recovered her new found power and dipped back onto the man's warm shoulder. She liked it here, safe forever and definitely in charge of him.

The years passed quickly, Jason enmeshed in his research, Kate busy with her work. All the while the little child grew, observing the seen and the unseen world around her.

The explosive cry of the hawk startled Jessica. She momentarily stopped playing with her dollhouse and watched the bird circle above, pleasuring itself in the slow spirals of a newly found thermal. Again, the hawk broke the still air—*scree ooo, scree ooo,* a gift of some other god somewhere in time, eerie, piercing, reverberating through the canyon searching for a kindred spirit, a playmate with which to soar.

Jessica watched, wishing she had wings like the hawk. She wished she had the awesome presence of the gray and brown lord of prey circling above. Her mind reached out to the hawk. Contentment with the doll in her hand vanished as did contact with the grass on which she played. Her young lungs sucked in cool air, her hair moved out of her face, brushed by a rush of cool air above the canyon. She was hungry. Her stomach thrilled with the sudden dive down, out of the thermal, eyes scanning the foliage below with a clarity she had never known, keen eyes looking for movement in the brush. Something small. Together they focused on a wee four-legged creature munching something as it foraged for its own meal.

Something new struck her, coldness, determination, no remorse, just hunger, and suddenly, fear, deep, chilling, fatal. In an instant she switched from predator to prey, not consciously but captured in another world. The thrill of flight abruptly changed to dismay. Too late, she threw the hand of her mind over the small furry four-legged creature now inches off the ground, moving with the swiftness of the hawk, the pain of talons gripping its helpless body. *Screeee-oooo, scree-ooooo, Screeam! Screeeamm!*

"Jessica! Jessica! What is it? Are you hurt?" Jason reached out to pick up his daughter as she ran toward him. His strong arms lifted her off the ground screaming and kicking, fighting with her small arms and fists. His nausea dispelled instantly, unaware she was still attached to the wee furry creature.

Then, disoriented, he blinked. Her dark T-shirt became a visual doorway, a window into deep space, through which stars from the universe rushed. The first sign. He looked into her eyes and felt himself falling into star-filled space and accelerating into the abyss of space. As his arms gathered her, folding her into himself, keeping her safe, he felt her passing through him, merging her thoughts and feelings with his. He felt her joy and anticipation, but his tears reflected his own sense of impending separation, leading to what, fear of the future? Loss?

She knew not that she was being lifted and carried swiftly toward the house nestled against the hill on the north side of the canyon, nor that her friend was screaming too, against what she did not know.

"Watch, OmRa. She becomes familiar with her gifts."
"She will learn more as she grows. Mahrianne has chosen well."

Young Jessica grew steadily more accustomed to her powers, but *now* she had a little brother to contend with. One afternoon she sat with four-year-old David on the back porch just out of the sun's reach. He ran his toy dump truck adding its sounds in as irritating a manner as he could design. Jessica ignored him with her superior, or so she thought, intellect. But she was annoyed, massively so. Her negative thoughts caused her to lose control of her favorite toy, and she levied her anger at the toy. As it flew at her, remaining just out of reach, she managed a swat at the irritating robot dragonfly. Its quick maneuver to a spot just beyond her fingertips annoyed her even more. She then swatted at David with equal success that then evened the score.

As a bonus, he taunted her with his favorite observation. "Yeah? Well, your Mother Earth is pissed!" This usually covered about everything and instigated some sort of mayhem, but not today. Jessica ignored him further and watched the spider web being constructed between the potted tree and a bush. When it was finished the spider disappeared into a shadow. She wondered where its *keeper* was. Then she quietly directed words to David's back, just audible, "I'm telling!"

That night David sat up wide-eyed, his little heart racing with excitement. Something had turned on his nightlight. He climbed out of bed as quietly as he could and tiptoed to the window by Jessica's bed, peering out into the patch of light covering their neighbor's yard. Even pushing his nose against the only windowpane he could see nothing different. But he knew they were there.

"Jessica! Jessica! Help me!"

"What is it David? It's almost four in the morning!" The bright characters on the face of the clock showed numerals three and forty quite clearly.

"I don't care! I can't reach! Hurry up, help me see out."

"Why?" But Jessica felt their presence also. "Shhhh. We don't want to frighten them." Then, with big-sister authority, "*If* anything is really there." But she knew they were. Quickly she fetched the dressing table stool and quietly pushed it up to the window.

David mounted the stool saying, "Shhhh, Jessica."

Together they peered into the luminous patch of light and waited. The few moments it took for their eyes to accustom to the dim far edge of the light patch seemed an eternity. David began to fidget. Another *Shhh!* from Jessica. The clock now said 3:42, but it seemed longer.

"There," David twitched his shoulder into his sister's side. Soft shadowy creatures moved slowly into the patch of light, eyes transfixed on something only they could see. First came a young trim mule deer followed hesitantly by her foal. They made no sound. They stopped transfixed, unmoving in the light. Slowly they moved forward, munching the grass as they passed. No more came. David and Jessica watched as the deer made their way across the drive and into the hill brush. Then they were gone. The two children hugged each other but said nothing. They didn't have to. They communicated in that other way.

The children turned as they heard movement behind them. "Mommy! Mommy!" cried David. "Guess what we saw." As usual, Kate had seen what the children had seen. Their minds were so tuned to each other that words weren't really necessary. She shared their excitement and thrill of discovery.

"He had big ears."

"She," corrected Jessica.

"SHE had big ears. So did the little one. Let me see your ears, mommy."
Kate laughed. Jessica rolled her eyes.

*OmRa and Thiesha watched the golden glow around the Indian Grandmother
and child change, now blue, now white, deep purple lightening to a glowing
lavender shade. The lavender glow turned and thickened, enveloping them in a
cloak of joy.*

*"The old woman passes her wisdom to the child in pictures, sounds and
visions," said OmRa. "She does not need words."*

"The child will be wise indeed."

"He will be taught his role as he matures."

*"Just prior to the end times," said Thiesha, "he will be led by the old man
down into the underground caves of Montezuma Wells from whence the Hopi
emerged after the last, the Fourth World, was destroyed ten thousand years ago."*

The infant in Grandmother's arms looked up at her, its eyes wide and
intelligent. They seemed large for such a small bundle, glowing bright with
an inner light of their own. A broad smile showed no teeth, maybe one tiny
point of white in pink gums. He wriggled with excitement, kicking his feet as
fast as he could. Baby hands tightly grasped each other then alternated with
attempts to reach her face. Chortles and sounds both delighted and startled
him in response to her soft words and small clicking animal sounds. Locked
in each other's souls their world seemed small and safe. Bright love flowed
between them, and from a distance the Ancient Ones watched, reveling in
the babe and his Grandmother.

She stood motionless looking at the immense owl quietly perched on the
side of the open window. Their eyes met gently, unblinking in the cool light
of the bright full Moon. Their spirits met somewhere in between, hers soft
and warm, the great owl's aware and soaring. Silently she gave permission,
as silently as he had asked it. No sounds were exchanged, simply a shared
knowing.

The new child stirred in his sleep offering a happy cooing sound as his
new mind and spirit joined his guide, spirit mentor and teacher for life on
its flight through the warm night. The Grandmother closed her eyes and
remained motionless as she joined the two on their flight through the night,
wonders and ways to be shared, other spirits to be met.

After several moons had passed, the small child heard the flute and drums and his feet began to move as had the feet of the big people. Side to side its little arms moved, back and forth over his head like the tree in the wind. He moved closer to the players and they turned toward him as they played. His mother moved closer, watching. Other children joined the magic child and moved in joy celebrating something they did not understand. But the old man playing the flute understood. This was the magic child of the Hopi. At the time of transition he would bring the Hopi's stone tablet piece from its sacred hiding place and place it together with the three missing pieces.

Now the dance changed. The adults saw what was transpiring and joined the old man with the flute. They began to move to the beat as more drums and flutes and rattles joined in. The dance isolated the magic child exuding such joy and awareness of its own importance that all exulted.

TaWa, bright spirit of the dawn, painted great tongues of orange and gold across the turquoise of the eastern sky, bright greetings to the desert creatures and the keepers of the secrets of the Red People, trustees of the future of all humankind. The piercing cry of the young great-hawk bore the heavy sadness of the human race, alone, lonely—*scree-oooo, scree-oooo*. Gray-brown wings riding free on the hot desert thermals, circled, dipping slightly. He allowed the searching sun to show the deep red of his tail feathers. Then, briefly, the sun illuminated the white fringe accenting the deepening red of his tail as he banked into another curl of the unseen spiraling air.

The young hawk rode the twisting thermal with the élan of a boy on his skateboard, screeching into the turns, adjusting to the changes of the thermal. As the thermal turned faster, the hawk turned, thrilling to the challenge, easily minutely adjusting the cant of his wings to maintain his place. The small boy at the edge of the high mesa chortled with glee and ran from his grandfather as fast as his newly found legs could manage. The quick hand of the boys' father interrupted his attack on the thermal before a disaster might occur.

Other birds found for themselves some leafy perch, respectful of the great hawk crying her opinion to the cloudy sky. Few flying things actually feared becoming her next meal, but while it was way past nesting time, respect and avoidance seemed prudent. Only crows or other great birds challenged her magnificent wingspan and reputation.

Her slow glide up the canyon echoed her cry. Those with astute hearing heard the faint answer to her cry from her mate yet another canyon away. A spiny yucca plant became her temporary resting place.

The small male hawk hovered eye to eye with the small male human, frozen in a bubble of time, buoyed by the hot desert thermal. The old warrior slowly moved the boy forward, silent, enmeshed in the endlessness of a warp in the fabric of space and time. And for a moment, neither existed—no time, no space, only the locked spirits of two small open minds. A representative of this one point in infinity, locking together from some point that never came, never went. A piece falling to its place in a script not created by man, but rather a soundless voice both from within the Great Mother and from the Great Spirit of the Hopi, keepers of all humankind.

The old warrior slowly removed his hands from his grandson. No watcher, no one waiting made a sound or moved an inch. The boy reached for the head of the hawk and the two floated in space, a space that was not space, in a time that was not time. All knew their time to fulfill their sacred trust had come. Their warm hearts, energy, and devotion to the will of the Great Mother would guide them.

Slowly the child drifted back to the mesa top. Chortling in glee he settled into Grandfather's open arms. They stayed there, frozen in the moment, looking deep into each other's eyes, and all else in time waited. The pitch of chanting and pounding drums heightened, joining the chorus of sound. Behind the faces of the Ancient Ones, clouds boiled gray and white. The dwellers of Black Mesa, the floating visage of the Ancient Ones, all watched the light around the boy swell to join that of the Grandfather. As they sat locked in each other's souls, a flow began. At first there was only a soft luminescence and sound from the air around the faces of the Ancient Ones, weaving a tapestry of sound and vibration in the air, growing, focusing on the old man and the boy.

Those gathered on the mesa top moved in rhythm, chanting, arms outstretched, deep primitive emotions swirling through them with no boundaries. A thread of light, indescribable in earthly hues, bright, luminescent, expanding, began to pass to and through the old man and into the wide-open transfixed deep black pools of the boy's eyes. As this thread of light grew, lightning crackled in the clouds, and all were joined in a powerful transfer of purpose from the Ancient Ones through the old man to the boy. And then, the peace of destiny as softly the boy slept in the warmth of his Grandfather's arms.

Joy huddled in small groups, awe and quiet pervaded, and the dwellers of Black Mesa began preparations that would guide and teach the boy. The tears of their ancestors freed from time's bondage lent voice empowering the child, winds of the past catching up to the now, rushing past to their destiny with humankind, the clouds of faces of countless waiting ancestors.

"Evening approaches, let us watch."
"He will find his powers. The boy's spirit guide has come."
The boy called to the swooping hawk, *hoo-ee-oo, swee-oo, swee-oo*, screeching as best he could in imitation of her cry. She paused in flight, adjusting the camber of her widespread wings, hovering in a thermal rising from the desert floor a thousand feet below. His eyes were fixed on hers as she hung almost motionless in space no more than a hundred feet from the edge of the mesa upon which the boy sat.

Come closer, his eyes commanded. Slowly he raised his small arms toward the majestic bird as if to embrace her.

Hoo – scree – oo, hoo – screeee – ooo. She flipped sideways exposing her under side to the boy. The sun silhouetting her transfixed forever in his soul the red of her tail, the white patches in each wing and the golden shimmer of the lowering sun on her feathers—her aura of pure gold and light. Suddenly she dove almost straight down, talons extended as if to snatch some unsuspecting furry thing from its burrow. Then, at the nadir of her dive, she bottomed her plunge in a sweeping circle, seeking to gain speed as she flew. *Hooscreeeeeoo*, she cried, *screeeee*. Swinging into an upward path she flew from the bottom of the dive up the front of the cliff to the face height of the boy.

Motionless, he stood at the edge of the mesa, arms outstretched, eyes unblinking. The old man once again put his flute to his mouth and softly blew magic into the air. Neither the hawk nor the boy turned to acknowledge him, but hung together, suspended in time, alive in another space, mind to mind, spirit to spirit. Both hovered in the shimmering glow birthed by the vanishing sun.

The old man's flute began a haunting flow of sound, ancient notes, links to the spirits of the past, awakening promises that grieved for fulfillment. He rose slowly from his cross-legged pose to full height in one smooth move as if once more lifted by the legs of youth. His feet slowly began stamping out a beat long forgotten, making patterns in the dust of the mesa top, and the mesa rang as a bell – pom – pom – pom –.

With each resonant vibration the boy and the hawk rose in unison ever slightly, more and more into the air. They seemed united by some invisible thread, some magic of the flute or perhaps a small anomaly in space and time. The old man knew the boy's spirit guide had come to claim him, to take him on his first journey, to teach him great things, to show him the ancient secrets of the mesa, and to forever bind their paths into one.

The boy's father stepped from the medicine lodge. What Gray Wolf saw was his son hanging there in the hot air, shimmering in the light, seemingly several feet off of the ground. Probably an optical illusion caused by the heat waves. Yet doubt invaded his mind and settled in the pit of his stomach already queasy from the strong tobacco of the lodge. He rubbed his eyes and turned away, glancing this way and that, trying to focus on real objects, things familiar and in their proper places, not hovering, weightless in some imagined patch of sun.

He felt the brush of soft air stroke the nape of his neck, fluffing the feather hanging from his headband. He felt his knees weaken a bit as something deep inside him whispered, *Turn around, Gray Wolf, see your son as he is.*

And slowly he turned, hoping all was normal and his small son would be safely on Mother Earth clasped by her gravity and not hanging in mid air over the void from which the mesa rose. That was not the case. The boy still hung happily, now facing away from the mesa apparently looking at the vast plain and other mesas both near and far. In his mind he called out to the boy, *Come back to the Earth, my son, for I know who you are now.* Slowly the youth drifted back to the mesa top and came to rest a few feet from his father.

"Father!" he cried. "I saw our canyon, the one Grandfather speaks of. I can go there. Can I Father, can I? You can come with me, let's go, let's go now!" he cried in childish delight. Gray Wolf's heart melted. He felt no fear, only deep ancient love for this small noisy boy, still slightly shimmering, accepting his newly found freedom, impatient to share it with his father.

"Come," said the boy. "Let's hold hands." Gray Wolf reached out and clasped the small dirty hand in his own strong man's hand. Slowly, ever so slowly, something passed from the boy into his body through the small hand. As it invaded his mind and his heart, the small face now by his side smiled up at him. "See, Father? Isn't this fun?"

"Yes, my son. But it is more than fun. It is a secret. In time you will see the gift you possess and you will share it with the deserving ones of our Mother Earth. But before then, keep it inside yourself, for you cannot yet tell with whom you can share your secret." He lifted his gaze, now somehow remote from his eyes, to the horizon and felt himself become light then lighter, unaware of his body or the feel of the mesa underfoot. Together their journey began. The Grandfather smiled quietly and watched.

The high-pitched cry of the young hawk broke the stillness of the warm afternoon. The urgency in the shrill *screeee-oo, screee-oo*, broke young Wind Runner's train of thought. As he scanned the sky searching for the noisemaker, a whoosh of air caused him to duck his head as a smallish brown and gray hawk sped from behind him, close enough to his cheek to move the air. His eyes followed it, then another bird passed close following the same path as the young hawk. This one was black, shiny black and a little larger than the hawk. Must be a crow, he thought. Obviously the crow was in hot pursuit of the fleeing hawk, noisily declaring his plight and doing his best to escape his tormentor. Wind Runner wondered why the crow was after him. He had not seen that before.

Then, as he watched the drama unfold before him, two hawks, both larger than the one pursued, left their lazy cruising in the high hot thermal rising from the desert floor. They dropped, wings folded in a fast straight-down dive toward the determined black bird and the frightened pursued young hawk. There was no mistaking their relationship with the young hawk. Clearly mom and dad were proclaiming their anger in rapid cries, both reassuring to the youngster and certifying intent to do mayhem on his tormentor. The pair braked their downward drop ending feet extended, wings cupped to break their fall, one on each side of the shimmering black knight. The battle that ensued rivaled the aerial dogfights of World War II and occupied all three adults. Junior turned and headed for the offered shelter of the cliff rocks amidst the racket of angered parents, diving, swooping, grabbing at the black knight. Junior continued to cry his indignation but remained protected in the rocks. The crow dodged and darted, effectively evading the angry hawks, finally lighting in the protection of a nearby sparse thicket. The hawks returned to their offspring, clearly irritated with him.

What did all that mean? Surely there was some lesson to be learned from such a display. He would ask Grandfather when he returned from the fields.

VII — RECOGNITION

The old desk stood at the apex of Jason's horseshoe-shaped work area. It competed starkly with the myriad complex computers, displays, wires, cables, speakers and the like that surrounded him but he didn't care. Here he could relax, focus, and contemplate the image on the monitor before him. His involvement in the confidential 'Project Y' had begun as a simple question he'd asked years ago when genetic research had just been recognized as the next bastion of opportunity. It had been his world for the last several years.

He stared at the image on the monitor, sensing that his research project was taking a new turn, but what? The image seemed to want to speak to him, to tell him something important, but he could only look at it so long before the faint flicker of the screen bothered his eyes. He blinked two or three times, rubbed his eyes and stretched back in his new ergonomic chair. That chair had cost North Western over twelve hundred dollars and had been worth every cent of it to him. Even the fight over the cost had been worth the aggravation. He could sit for hours at his work now, no more of the fatigue and back spasms he used to suffer.

He peered over his computer to check the old school clock on the wall. A brown hexagonal frame surrounded its face painted with Roman numerals. Behind its glass face a pendulum swung, ticking off time. Back and forth, tick, tock, tick, tock. The incessant tick, tock never bothered him but it drove his assistant wild, so she stayed in the lab a good part of the day. Jason watched the hands. Somewhere in some crevasse in his mind, a connection began to form between the hands on the face of the clock and the marks in the image on his screen. The clock and the screen were directly in line, clock above screen. On the screen was a complex DNA pattern, different from any he had come across before. The differences were subtle but real. Anyway, it might have nothing to do with the project on which he worked. His mind filed it away.

Some weeks ago seven-year-old Jessica had paid him a visit at the lab. Kate had brought the kids to see where dad worked and then taken them to lunch in the cafeteria. Jessica had wanted to see what she was made of . . . *the stuff people are made of,* she'd said. She had noticed the pattern differences immediately, then reluctantly donated a small scrap of skin to the effort.

At home that evening the hot heavy air sat close to the land, still, thick, oppressive, even in this canyon so near the coast. Jason couldn't concentrate. He'd come close to recognizing something about DNA that had been nagging him, close to the surface of the fertile pond of his mind, but not quite close enough for him to grasp. The mass of air clogged his brain. He tried to shake off the lethargy but the struggle to rise above it had been fruitless. His pen slipped unnoticed from his hand. The sound of it bouncing on the bright spotless yellow oak floor too went unnoticed. Something, somewhere had taken over his thoughts.

An odd shroud of disbelief crept into his mind. He understood well enough the tricks the mind can play should some pre-learned or previous bent of the brain become challenged by the new, but he wasn't yet aware of the change. He didn't want to challenge all he believed about DNA and its amendable structure, nor did he want to address the queasy feeling in the pit of his stomach that usually came when new information challenged his accepted tenets.

Lost in thought Jason sauntered out of his book-lined study ignoring the strings of DNA molecules neatly drawn on the electronic whiteboard's smooth surface. He plunged his hands deep into his jean pockets and absently edged the glass door open with his elbow, wide enough for him to step out onto the shaded patio.

He and Kate had spent many hours with their friends Greg and Sara who talked openly about their life experiences and their unusual abilities... sometimes bizarre, from Jason's perspective. Kate shared some of the same extraordinary abilities. Jason found the whole thing unsettling and confusing but utterly fascinating. These divergences in their abilities surpassed those of most people yet could be found in varying degrees around the world.

Jason walked slowly around the patio thinking only of the potted plants artfully arranged about its edges. The hibiscus could use a trim, maybe some plant food. He nudged it with the tip of his Birkenstock. There were dirt and leaves in the corner by the avocado tree. Jason ambled down the path of neatly positioned railroad ties boxed into steps by brick and concrete. The steps descended the short distance from the patio to a small grassy area where he could hear the children playing. Two summers ago he had set up a picnic table with benches, a barbecue and small storage area for the barbeque utensils. He had trimmed the native brush into a respectable quiet enclave just above the canyon floor.

It was cooler here. Two huge pines provided a motley shade capturing what movement there was in the air. They had never added air conditioning to their house and seldom felt the need for it. The canyon always captured the sea breezes channeled in from the ocean, up through the basin and on over the Santa Monica Mountains—one of the nice things about living on the west side of Los Angeles.

Jessica and her friend Tracy were playing a game of *house* in the grassy patch while David played close by. The two girls played noisily together, having established house boundaries by outlining rooms with rocks. The grass provided the carpeting of the rooms. Jason marveled at the patterns of light that filtered down to them playing on their lightly clad forms. They looked oddly like light patterns rather than kids.

"Hi, daddy! Want to play?"

"Too hot. I'll watch, okay?"

"Okay." They were quickly reabsorbed in their complex game of house. He mused that parents could learn a lot by watching them play it. How clearly the mommy and daddy roles in the game pointed out his and Kate's own parenting flaws! He would have to discuss that with her later. For now he just sat watching the light patterns. The light reflecting off of Tracy's almost alabaster skin was bright and cool. The light reflecting from his daughter's was a rich warm mocha. He sat reveling in the beauty of his own two children, Jessica and David. They were a happy blend of the lighter beauty of his wife and the deep chocolate of his own.

Jason found himself relaxing, melting into the moment, surrendering to some swirling vortex of non-words buried deeply in his brain. Mixing. Moving. Reaching. Racing. Suddenly, like words on an electronic whiteboard flashing in bright neon the truth of the children, the difference between the play of his two alone and that of his daughter and her friend! He lay back against the table, his head spinning with too many thoughts, too many emotions. Excitement and fear surrounded him and nausea threatened to overcome him. He turned to the table and rested his head on his folded arms.

Slowly he began to recover. He waited for his legs to feel steady enough to get back up the small incline to the house, to a phone, to his wife. Then, like a blast, recognition hit him. *Oh, my God! Kate! She's just like the kids!*

Jason paced around the lab trying to explain his discovery to his friend.

"Crop? Crop?" Greg's response was uncharacteristically loud. "That's pretty subjective for a scientist! Do you really mean *crop?*"

"Yes, Greg! This *crop* of kids has been altered genetically!" Jason had been pacing around in the lab waiting to share this discovery with his friend. He'd expected a calmer response.

"Are you saying *all* of the kids have been altered some way? That's impossible! I know about my son and your two kids, but surely not all children!"

"I'm not saying all of them have been changed but many have. They know things, see things and hear things. They talk to people all over the world mind to mind. Not all of them, of course, but this must've been going on for a long time. It certainly would explain a lot of this world's mysteries and miracles. Kate knows things, too. She can communicate nonverbally with the kids. Sometimes it's as if she came from another planet." He ran his fingers through his hair.

"The kids play hide and seek in other dimensions. They told me this. They pop in and out of sight, teasing me. They can go just about anywhere. It scares the heck out of me. They also seem immune to danger. They don't seek it, but they seem to step sideways into a different time or space and return after the danger's past. They even come out of serious mayhem unharmed." Jason's shoulders slumped. "I feel like everyone's more advanced than I am."

"Well, what does your research show? Can you substantiate your hypothesis?"

"I think so, Greg. I've looked at their DNA. It's different from mine and maybe yours. But what do I do with this information? If I bring it out in the open and there's hostile criticism my whole research project could be scrapped! I'm not sure I'm willing to take that chance."

"How do you feel in your gut, Jason? What does your intellect tell you to do?"

"God, I don't know. The bigger question is, what does it all mean? Why are the children being affected this way? When these kids are in their teens or twenties are they going to change the world as we know it? And if this continues, are we moving toward some monumental event?" He dropped into his chair and leaned on the desk. "Maybe I'll just keep it under wraps until I'm absolutely certain. People just aren't ready for this yet."

"Well, good buddy, keep gathering your facts, then when you're ready, bring it out to the widest possible audience. Present your evidence, and let the cards fall where they may."

Kate dropped by the lab that afternoon to talk to Linda. She and Jason had discussed his theory long into the night. She was anxious to see for herself what it was all about. She stood at Linda's side looking at the lab monitor, disbelieving what she saw and only dimly comprehending its meaning. She knew she was different, she just didn't know exactly how different she was, or why. At first, staring at the representation of her own DNA had been simply curiosity. Now, with Jason's lab assistant pointing out and explaining details, her attention had snapped back to textbooks rife with information on the function of various segments and patterns. But the thing that jumped out at her at this moment made her question her instincts. She could only marginally accept what was happening.

Jason and Greg, returning from lunch, entered the lab engrossed in conversation. Linda remained quiet, sending a glance to Kate to do likewise so they could follow Jason's words. "The lab says it may be a mutation they haven't seen before. They're not even sure it is a mutation. At the allele associated with a specific segment in the thalamus this change seems to extend an arrangement of a series of loops in the DNA segment. We hadn't seen this before and assumed it was an unused segment." He looked up and acknowledged the women.

"Is it a change in the thalamus or something added to existing DNA?" asked Greg. "One of the traits apparent in people such as Kate and the children, as well as in my whole family, is heightened awareness. That includes being able to communicate with others like themselves without actually speaking."

Irritably Jason interrupted. "I can't speak at all without talking!"

"Then you may not have that particular trait because you don't have that change. Knowing you for so long, Jason, I'd lay odds on you having some genetic changes even though they may not be the same. Maybe there are even layers of changes acquired over long stretches of time and passed down through generations. Take a closer look at your family, my friend. You may assume that they're just very bright whereas they may be really very different!"

"I sent a sample from myself for analysis." Jason looked up at Greg perched on the high stool near Linda. Feeling like his mouth was full of cotton, he said, "I have one of those loops, but only one. Not several like the kids." He looked over at Kate, trying to hide the concern eating at his insides. "My guess is that they may have at least one loop greater than Kate, but probably more. And . . . that she has a hell of a lot more than I do."

The two men sat in silence. Kate and Linda just looked at each other. Jason's mind was a mess of confusion and dread. Greg was rapidly correlating pieces of what he had observed of Jason's kids with what he knew of his own aberrant traits. He broke the silence. "What you're implying is that there's another influence on the history of humankind, some extra world force messing with our genome."

"I think that whatever evolution is, it senses what we now face living on this iffy planet and it's trying to beat the odds against us. Something's on our side."

Linda raised her head but didn't interrupt. She had Kate's attention and was glad someone would listen to her. She had some thoughts on the subject herself! The conversation moved on, postulating the ideas put forth and finding more questions than answers.

Finally Kate spoke up. "Jason, you have to gather leading scientists and tell them what you've discovered. Give a speech to the planetary scientists and other key people. Some won't know the steps through the maze of DNA-speak so you'll have to make it crisp and clear. They'll go home and dissect everything you say anyway."

She and Linda quietly continued their conversation and left the men to theirs. "I'll explain the definitions," said Linda. She looked over at the men, deeply engrossed in their conversation with each other. "Let's go to the cafeteria and get some coffee. We can talk more there." Kate nodded her agreement and softly closed the door behind them.

As they walked, Linda began. "I'll do a top-down view. Stop me any time you want more information. Genotype is the genetic endowment or constitution of an organism. Phenotype is an observed trait of an organism. That refers to its observable properties, including the visible traits. A hereditary determinant of a trait is called a gene, which consists of DNA. The different forms of a particular gene are called alleles. Genes come in pairs. They separate in gametes and join randomly in fertilization thus introducing the element of chance . . . "

The men continued their conversation after Kate and Linda left. "Is the change temporary or permanent?" asked Greg.

"It's permanent, part of the genetic makeup from now on." Jason didn't know where the conversation would lead. The odd feeling in his stomach was much like the guilty feeling of a child being caught with something you weren't supposed to have.

"How long has this been going on?"

"Who knows? Our genetic makeup has been changing since humans first appeared on the Earth. It probably will never stop changing unless something drastic happens."

"Like what?"

"Like total annihilation or something close to it. Something akin to what you said—what the Native American Indians believe, that the world has been destroyed four times already and we're working on destroying it for a fifth time. And that's the last chance we get for a place to live."

Greg turned back to the issue at hand. "Let me play Devil's Advocate. So you discovered both your kids are different. They seem like nice kids. Why did you look at their cells under that advanced microscope? I would say they're smarter than a lot of kids but not very different."

Jason paused, thinking how best to share what he knew with his friend. Greg had as much technical background as he did, but in geology instead of genetics. He leaned wearily on the table and sucked his lungs full of air. God, he was exhausted! He looked up connecting with a quizzical but interested Greg and weighed the level of trust they shared. He was tired, but before long everything would be out in the open anyway.

"What is it old buddy? What's gotten into you? Why are you all bunched up like that?"

"How much do you know, Greg? Logic dictates that all of these changes, coupled with the forming light rings, means something big is near, something beyond Man's understanding."

"As a geologist I know that Earth changes are accelerating. Because of my background I'm heavily involved—as you are—with Civil Defense activities. I also know there's an increasing number of people who seem to be very smart but are also very different. Everyone acknowledges these things. You brought up the Indian's Fifth World myth. That's just a myth, so what are you driving at?"

"I don't believe the Fifth World story is *just a myth*. Anything that's been passed down through time and remains basically the same is based on something that really did occur. Unfortunately, it's Man's arrogant folly to dismiss these tales." Jason stood and stretched. "Let's get together for dinner soon. We can talk all night then if we want. Right now I'm beat."

Greg wasn't quite ready to quit. "Just one more thing, Jason. Tell me more about the kids."

"Okay." Reluctantly he continued. "You know my family well. They have traits that are different than other people. They perceive things I would never notice. They can do telekinesis. They play games with friends somewhere else, even with those who speak a different language.

"They know when a friend's in danger . . . the kids more so than Kate. I even suspect Kate's circle of female friends is altered. It must've been going on for centuries because the changes would be introduced carefully and experimentally. Just think how medical and societal changes occur gradually over time so as not to alarm the people. This modification would be introduced in a way that doesn't create fear or start off another inquisition. People are still largely what I might call primitive, necessitating the gradual introduction of significant changes. Great leaps in science often occur from the work of special people like Einstein, Galileo, Jesus and many more. Genetic research is going on today at an almost frantic pace."

"Things are accelerating," mused Greg. "They may even be related to the apparent speed up of Earth changes. The magnetic field of our planet is changing and might cause another flip of Earth's poles. That's happened several times before in Earth's oral and written history with disastrous consequences on Earth's surface. There are references in Chinese history, the Bible and so on, tales told word of mouth, generation after generation, ancient myths. These all try to account for the enormous changes through time as best they can."

"You're right, Greg. In spite of all the gyrations this planet makes, humankind clings to the surface no matter the struggle or consequences. The Life Force drive we have must appear overwhelming to any other civilization that would be able to recognize our struggle here, whatever its nature, wherever it's from in the universe. You'd think we would get their attention, at least to determine if we were worthy of helping or even if we could be helped."

"Everything gets blamed on the *gods* or *divine intervention*," said Greg. "Both imply that events aren't understood by the inhabitants of our planet. If it were true, then surely life here would be the extreme sport of the gods.

As the saying goes, 'They put us here with no money and no skills. Go ahead. Have fun!'"

That brought a chuckle from Jason. "At any rate, today an awareness of massive changes is creeping into our lives. Many people choose to change, and many choose to ignore. But whichever it is, it's here with us, now, today. So I don't quite know what to make of this, but it scares me to the depth of my soul. And I do believe we have been modified, carefully, for an untold number of generations. Carefully, because I believe humankind or our precursors' *helpers* may have been far ahead of us in evolution. Just look at the world myths. The most pure and pre-sage myth is the World Covenant with our Maker carried through time by the Hopi Nation. You told me that yourself, Greg. I'd like to hear about your knowledge of the Hopi sometime."

"I'll tell you what I know of it. Let's plan dinner soon. I'll ask Sara to call Kate and pick a time, okay?"

"Okay, my friend. Sorry I ran on so long, but this has been deeply on my mind now for quite a while."

VIII — MOTHER EARTH

"He is hesitant, Thiesha, but he is the one to spread the message."
"His friend will be the catalyst. Have patience, OmRa. He is a scientist. He will find his own way."

That evening, the two men sat quietly at the dinner table, relaxing and discussing their work. Anxious to hear about Greg's recent study trip, Sara and Kate hurried to remove the remains of a simple, late-evening meal that had not included the kids. All three children were happily sharing the sleeping bed Greg had built for Robbie. Sara paused as Kate put the last dish in the dishwasher. "It's Friday and the kids can sleep in tomorrow. I told them Jessica could have the couch and the boys could both fit into the *structure*, but Jessica insisted on joining them so I left it for them to figure out. They're quiet, so I'll assume they're okay."

The *structure*, as Robbie's bed had been dubbed, consisted of a gymnasium-type frame of heavy plastic, amply padded and decorated in the fashion of commercial play areas. The kids loved it as did the moms and dads who could now enjoy a quiet evening together.

Sara flipped on the switch to the coffee maker she'd set up long before anyone had even thought of dinner. "Shall we?" she asked, her warm smile beckoning Kate to follow her. She picked up a plate bearing scrumptious after-dinner pastries and Kate gathered the mugs to follow her to their comfortable living room.

"Come on you guys," Sara called. "You'll be more comfortable in here. Coffee will be ready in a minute." She set their goodies within easy access from the couch and the two over-stuffed armchairs.

When all were comfortably seated Sara asked her husband to share what he had learned on his Volcano Lab trip up to Alaska. "I'd like to hear more about it, too," she said.

They'd all heard bits and pieces since he returned and now they were anxious to hear the rest. Greg was excited about the findings but also troubled at what he'd learned. They talked for a few minutes as the coffee brewed, its fragrance setting a stage of peace despite the pending topic. Soon each had a steaming mug and Greg began his tale.

"This trip fit perfectly with my work as a geologist as well as at the Civil Defense Authority. If the new information is correct, and it sounds like it is, it'll be a big step forward in disaster avoidance. Our team traveled to North America's famous volcano on the edge of the Gulf of Alaska, named appropriately *Pavlov*. It's erupted forty times in a little over two hundred years making it North America's most active volcano. We don't hear much about it since it's thirty miles from the nearest community and doesn't present much of a threat to anyone. But it's become a lab for our struggle to predict when a volcano will erupt. The object, of course, is to save neighboring areas from death and destruction.

"What our lab observed is that Pavlov's eruptions follow a pattern. It doesn't erupt every year, but its periods of activity always seem to occur in winter or fall months. Twelve of its sixteen eruptions since 1973 were in the autumn."

"Sounds like you're onto something, Greg." Jason had worked with him on Civil Defense projects since they first met. "Anything that saves lives is important."

"Even more interesting," continued Greg, "is that four of these occurred within the same five-day November window, but in different years. This phenomenon has been observed as well in other volcanoes. For example, Japan's Sakura-jims volcano has a strong tendency to erupt in December, January or February."

"Is this phenomenon a worldwide tendency?" asked Kate.

"Many of North America's volcanoes show a preference for seasonal eruptions, and nowadays as more and more people live closer to volcanoes we have to be more precise in our forecasts."

"Here on the West Coast we worry about earthquakes first and foremost," said Jason. "Next, we worry about volcanoes. We have only some minor angst about hidden faults or unstable land areas unless we live on or near one."

"What causes certain volcanoes to erupt in fall months?" asked Kate.

"For centuries researchers tried to come up with numbers and observations that would help predict and prepare for eruptions. Today, however, a simple look at the problem shows that some volcanoes build up pressure toward bursting, and this pressure seems to be related to stresses caused by changes in the weight of ice or water against the walls of these volcanoes." He paused to add some cream to his coffee.

"I've heard that the Yellowstone National Park caldera is slowly rising," said Sara. "That's an old volcano, but it wouldn't be affected by water, would it?"

"You're right. There are other stresses that cause pressure and increase the probability of an eruption. If you want to get down into the depths of the geology, earth movement is one of the biggest factors affecting our lives. Yellowstone is a broad plateau in the northwestern corner of Wyoming, the very heart of the northern Rocky Mountains. It has the world's largest volcano and there's evidence that it's had at least three horrendous eruptions. More than six hundred thousand years ago a massive volcanic eruption wiped out an entire mountain range that ran through what's now Yellowstone. The thirty-mile wide crater is the size of the Los Angeles Basin. It left a mountain-encircled plateau that was subsequently changed when a collapse formed the huge caldera. Old Faithful, the most popular geyser in the park, was named for its regular timing. But recently the time between eruptions has been disrupted by changes in the earth underlying it. Events now occurring indicate a more active state for the entire park."

"Hmmm, I don't think I want to go there for vacation!" said Kate.

"The Yellowstone volcano is a geologic hotspot of molten lava, or magma. These volcanoes are found around the world and mostly originate underlying the thin crust of the ocean floors. The hotspots that underlie continents, however, are rare as Earth's crust is over twenty-five miles thick. When they do exist this causes a stationary plume of magma to rise up through Earth's mantle into the crust forming a shallow chamber of magma near the surface."

"Oh, Greg, you sound like the professor you are," said Sara kidding her husband fondly.

Greg smiled and continued. "As the continental crust moves southwest, the ground above the stationary magma chamber stretches and thins and expands across an wide area of land. The growing magma chamber is continually being fed magma from the plume. As the ground thins, more and more magma accumulates and causes a bulge in the crust. In the past, that magma grew enough to weaken the ground over the hotspot that then erupted onto the surface. This caused the largest volcanic eruption ever recorded. Eventually the hotspot moved farther northward as the Yellowstone volcano moved away, setting the stage for its next eruption and leaving behind another deep caldera in Yellowstone. The three eruptions have each left calderas that are visible both from satellite images and from observation points in Yellowstone Park."

"Are we due for another eruption there?" asked Jason.

"The most recent eruption delivered a blast one thousand times more powerful than Mount Saint Helen's 1980 eruption. Will we have another eruption soon? It's been over six hundred thousand years since the last one. Benchmarks since 1985 reveal uplift of about one-half inch per year associated with rising magma. There's also been a period of subsidence but the meaning isn't clear. Should one of these magma chambers blow, the threat could spread enough dust and gas to create worldwide disruption."

Jason added a little brandy to his coffee. "I think I'll enjoy life as long as I can."

"It's a marvelous place to visit," said Greg. "Early explorers opened the world to this massive place and its prodigious yellow stone from which it gets its name. We know its marvels through our vacations or from our classrooms. But there's a dark side to this magnificent Eden. An enormous volcanic plateau about eight thousand feet high dominates Yellowstone and extends over a thousand square miles. The old rolling lava flows are easily spotted. Yellowstone volcano has its own cycle of life. In human time, it's a very long cycle. The introduction of satellite reporting on Earth has made it possible for us to see just how widespread the devastation from its rare eruptions has been. It erupted three million years ago, then two million years ago and again about six hundred thousand years ago. Each of these eruptions has devastated life on Earth leading geologists to point out that the Yellowstone volcano may be overdue—not a happy prospect, particularly since the Yellowstone calderas match the world's largest in size."

"Let's hope it doesn't happen in our lifetime," said Kate. "Could Civil Defense store enough food and water to tide humanity over until the atmosphere cleared up?"

Jason met Greg's glance. "We're trying," he said, "but it's a multi-year job. We've got stockpiles stationed around the city and we've got trained volunteers to oversee the distribution but it will depend on the cooperation of the public to keep down hysteria. Let's hope we have enough time to prepare before anything happens, whether it's a volcano or something else."

"The Indian myth of the Five Worlds contains several references to a worldwide catastrophe," said Greg. "I'd like to explore that particular myth more closely."

"What's it about?" asked Kate.

"Well, in a capsule, it tells of the past destruction of the first Three Worlds, the coming destruction of the Fourth, which is our world, and the opportunity for one last world, the Fifth. It tells of a sacred covenant with the Hopi Indian Nation for them as the guardians of that opportunity, for that one last try to get it right in a Fifth World, the promised new race of humankind and a rebirth of life in peace."

"Wow! That's some myth," said Sara. "Where would one begin to investigate?"

"I'm planning a trip to Black Mesa in Arizona," said Greg. "That's the seat of the Hopi myth."

Sara raised her eyebrows at that. She wasn't the only one surprised. "When were you planning to go?"

"Soon. I've studied the mesas for a long time, but not only the geology. I'm as fascinated by Hopi folklore as I am by the geology. I have a strong feeling I'm missing something," he said softly. "Ever since I returned from Alaska I feel like I'm being pulled toward something. Black Mesa is a rugged area. I'm not even sure how I'm going to approach it."

Jason shifted uncomfortably in his chair. "Look, Greg, I'm not so sure it's a good idea to go alone. I could come with you."

"No, this is something I have to do on my own, but thanks." At that he launched again into his evening treatise on volcanoes. "As I was saying, one example of earth movement affecting volcanoes comes from the western boundary of the South Pacific. The entire Pacific Plate is part of what's called the *Ring of Fire*. In that subduction zone the Pacific Plate breaks up at the southwestern segment of the Ring of Fire and it is . . ."

The discussion among the four continued late into the night, far beyond the last cup of coffee. Jason and Kate finally said their goodbyes promising to pick up the children in the morning.

Greg bounded into Jason's office a few days later eager to share his most recent calculation. Their discussions of geology and earth movement had sparked a lively dialog between the two. It had also initiated a new line of research on Greg's part that now caused him a frisson of alarm.

"An *inch*?" Jason laughed, slapping his knee as he stomped his foot in mirth.

"What's so funny, Jason?"

"Excuse me, but a piece of dirt? Aaah, hah, hah, HAAAAaaa." Jason practically choked as he banged his knee with his fist. "A piece of *dirt* moves an inch in two weeks and you're upset?"

Jason's guffawing and knee slapping irritated Greg. "Well, let's see now. What would a one-inch movement in a slab of dirt be capable of doing? Assume this slab is part of a mountain, which happens to be an island. What if that one-inch is only the beginning of a slip into the sea? And what additional dirt might it take with it into the sea?" He waited a moment to see whether he was beginning to get Jason's attention.

"The crack extends from sea level to a point on the side of the mountain about four thousand feet above sea level. There it slopes down to sea level twenty-two and a half miles from where it began. If we stand on the side of the mountain at the highest point above sea level, the slab looks like a wedge-shaped piece of pie. The ground slopes down to the sea, so we make assumptions about how thick the pie is."

"Is this a long story?"

Undeterred, Greg continued. "For example, assume the average height above sea level is one thousand feet, and the average density of material in the wedge weighs about that of sea water, we would be looking at a mass of four point eight million tons. If we assume an average height above sea level is two thousand feet, we'd be looking at a mass of nine point seven million tons. The greater we assume an average height to be, the greater the weight of the mass will be.

"If the chunk we first looked at includes additional mass that breaks off under water, we can add the mass of the part under water. At some point it's all going to slide into the Pacific Ocean from the slope of Mauna Kea in one big rush." Greg stopped talking, waiting for the numbers to click in and waiting to let his words penetrate his friend's mind, an irrepressible smirk playing about his lips. "So, what, exactly, do you think is going to happen?"

Jason had stopped laughing and wiped his tears with a tissue. The quiet that ensued was like a vacuum sucking the air out of the room. "Oh, Lord!!" Jason's face went ashen.

Greg chided, "Perhaps a tsunami one mile or more high headed for the West Coast of the Americas at two hundred miles per hour? Stop trying to do the math, Sparkie, you're just a layman. You'll hurt your head."

"Stop with the arithmetic! I get it already!" Jason tossed his pencil at Greg.

"Okay, pal. It's just that events are accelerating and I'm getting worried. I'm finally going to make that trip to Black Mesa. I'll either get some answers or simply ease my mind that it's nothing."

"I'm concerned, too, Greg, but not only with your Earth events. I feel like the children are drifting away. They're normal when I talk to them about day-to-day things but then they focus outward. Their minds are in another time or space. I don't even know what I mean. It just scares me, that's all. Even Kate..."

"What about Kate?"

"She's quieter, more serene. She doesn't acknowledge a change, but I feel it."

"Maybe you're imagining it. We're doing all we can here on the home front to be ready for whatever happens. Just hang in there."

"I need to do something, anything!"

"I agree. You continue your DNA research and I'll go after the Five Worlds myth. Maybe the two of us will come up with something."

IX — BLACK MESA

The old Grandfather stood abruptly awakened from his nap by the call of the west wind. A shiver ran down his spine and back up to the base of his skull. He knew without thought that now was the time awaited by the Hopi for execution of their sacred duty to humankind. His strong legs trembled slightly as if a warrior's call to duty from some ancient horn had been sounded. Quietly, slowly, he turned once to face each of the four directions, then stood facing west. Silent and still he called to his spirit guide to locate Greg. Thus he answered the first phase of the responsibility of the Hopi.

Not long thereafter, White Feather sat alone in the early morning light facing east, waiting for the Sun to bring her bright warmth to Black Mesa. He counted the number and kind of animal spirits she had displayed for his pleasure. If there was any pain or discomfort in his old legs as he sat cross-legged on the hard black stone of the mesa, he did not show it. He waited, then heard the soft pad of moccasins on smooth stone, silent to ears other than his own. Without turning, he gestured toward the magnificent sunrise display being prepared for him. *Ah! Both eagle and wolf clouds today! Someone comes. He is expected and time grows short.* White Feather's lined face broke into a knowing grin. Rising gracefully from his perch of black stone, he turned quickly, a broad smile now on his face for his son.

Yes, we have waited long. Gray Wolf spoke to his father mind to mind. *Your council has been wise. I will gather the Elders. I think they know.* He shook his head slightly, returning his father's smile. How the old man could hear him as he played this silent game he could not quite fathom. His father behaved as a man far younger than his years, but even so, he had eyes in the back of his head.

Gray Wolf moved through the village waking those needed and returned to the place where White Feather waited, still watching the sunrise. White Feather turned and motioned them closer, then nodded to Gray Wolf to quiet the village. Silence came swiftly as the much-respected Old Ones appeared. Their responsibility to maintain the tribal charter of the Fire Clan was to be concluded before their passing. Too many of the younger members were occupied with other pursuits, and time to carry on the work of the Fire Clan would not seem so urgent. The tribal Elders knew their parts well. They were thrilled and anxious to begin preparations for their fulfillment.

Gray Wolf joined his father near the edge of the mesa. Both sat cross-legged facing the village Elders who settled themselves, also sitting cross-legged on the ground in spite of their age. The remainder of the village population gathered round but maintained some measure of distance from the seated Elders, sitting or standing as their wont. Together they waited.

The road stretched across the desert flatness for miles, scratches in the soil of the ground, Earth's morning blood on the ground, claw marks from an enormous monster. Carlos Santana's *Evil Ways* played on the radio of Greg's Jeep. He turned the volume high, full right, pouring the throbbing music over him. The vibration penetrated his bones and blood. He pushed back against the padded backrest, relaxing into his mind, turning his body over to the visceral music and the road.

Relaxation deepened and he became one with the tide of sound, stepping through a curtain to walk among the stars. The flesh of his weathered face drew tight pulled by some involuntary tightening of the scalp. His head tilted back, features wolf-like, eyes slits in a tanned and leathered face—a hunting animal in tune with something beyond what could be physically sensed, beyond an awareness of the situation.

Black Mesa rose from the hot sands of the desert floor, glowering, majestic, Chief of Mesas, Keeper of the Earth. Rivulets from the last rain gleamed red as their waters sought something lost in the sands of the rocky plains. Distant clouds roiled slowly, a living backdrop for the mesa, lashing out at the encroaching blast of the mid-morning sun. These, accompanied by loud thunder and bravado, punctuated by muted streaks of lightning, jagged spears angrily holding back forces unseen but felt, white man's greed, savage lust for the black heart of the looming mesa.

Greg stopped and stepped out of his Jeep. Forty-seven years had taught him more than he wanted to know, not just about the science of geology, his chosen field, but also about the fringes of science—studies of the abnormal, the paranormal, and angels. But the desert had long been of deep interest to him, particularly the mythologies of the desert tribes. Myths handed down by the spoken words of tribal elders, wise men and women descended from ancestors from the beginning of countable time. In surprise he felt a pressure weigh slightly on his shoulders. He'd thought himself trim for his six-foot height and his ability to compete with hot kids in the cool California surf.

The drone came slowly from all sides, pulsing, now soft, now strong, a beat emerging through the monotony of the passing day, subtle variations in sound, a living, breathing being of many mouths. There were sounds of brush animals and birds, deep, primitive, almost primordial, a gift of ancient hands and souls talking in their own language, speaking to another spirit, another soul's vibration. Penetrating, changing that which it touched, penetrating, speaking of wants. I want. You want. A chant of soul so ancient that words were of no count, mind-numbing yet freeing. And through this a chimera grew, a vast sad but threatening cloud kachina.

Greg drew a slow breath and let control slip away. At some primitive level he knew he was summoned and under other rules of reality. Beside the kachina Greg could discern a faint image developing—a female of some sort, not much like women he knew, but wispy, red hair billowing in some unfelt breeze, smiling safety and welcome to him. He dropped the last vestments of control and waited.

He stood peering at the magnificent spectacle of the mesa before him. The apparition seemed more of an entity than part of the land. A primal thrill sped through his man's body shaking it into awareness. The mesa was speaking. He stood frozen as some heightened sense overtook him. The voice of the mesa spoke to him echoing down the stairs of the trail and reverberating like a pinball amongst the many mesas of the desert.

He had no concept of time or of how long he had been there. He and the mesa were connected. In his mind he reached out seeking contact. But the manse of the mesa was too powerful. He waited, not yet fully seeing the figure suspended between him and the mesa, becoming clearer, hands outstretched to him. As hands grasped hands, a surge of some ancient energy passed between them, uniting them as one purpose, one mind. Together they ascended swiftly, lifted from the floor of the desert onto the flat surface of the mesa by an unseen hand.

The hand now holding Greg's was different, rough and wrinkled, worn with time and life. It was that of the Keeper of the Truth, Grandfather of Time, Speaker of the Secrets, and Keeper of the Covenant of the Tablets. Bright coal-black eyes met Greg's. With a voice emanating from the eons the apparition began to speak.

"As I stand tall with the Ancients, I hear the sound time weaves. I am at peace and I feel the sound of truth. The message is borne on the leaves and the winds, it never ends, never begins. I listen not to the words of men for they are only mirrors of their own arrogance and ego. The tide slips. The tree stands. The Moon glows, and this I know, peace is truth and peace *is*.

"Wrongs of the present do not correct wrongs of the past. In the future, wrongs of the present become wrongs of the past. With this logic the future will have no sanity and know no peace. Stand with the Ancients and listen to the sound of truth; discover peace.

"In the coming storm, people, as the leaves, will only share their flights along the jagged edge. As the mindless shadow of the rocks, people will exist exposed to the whims of fear, furiously pumping their legs toward oblivion."

On the mesa, members of the Hopi village had gathered—these, the ones who would bear witness to the end of the Fourth World and initiation of the Fifth, the watchers and keepers of their sacred trust. Old, tired, sad, hopeful to have lived long enough, this last generation, now knew they would fulfill their covenant with the Great Spirit.

Carefully, Grandfather picked up the wind flute. He sat cross-legged, face turned to the sky, and gently placed the flute to his lips, caressing it softly, blending with it. The caresses, the witnesses felt, were as if he were caressing a new born child or a warm woman.

Black Mesa lay silent, more than a thousand years silent. No sound could be heard. The flutist slowly drew a deep breath, a sigh from the wind, and placed his fingers fluidly over the openings of the flute. Then he began to play, not so much as in playing an instrument, but as tuning into an unheard symphony somewhere in its stream of the infinite, a rider on its haunting unearthly melody.

Shivers of ecstasy crept over the gathered clan members, rising from somewhere deep within each, yet connected to the unheard music of the infinite by the solitary flute player. The haunting train that flowed from the instrument was punctuated by bird-like throaty sounds as he drew each new breath and again took up the passing of breath through the wind flute. These haunting, almost mournful sounds touched ancient chords hidden in each of the listeners, unleashing deep emotions of caring and longing.

Silent tears crept down the flutist's cheeks from eyes closed in communion. Each of the gathered, the members of the clan, the watchers, faces upturned, wept silently, feeling deep, great compassion for humankind and a burning desire to reconnect with the soul of the universe and the design of the infinite.

The chant began, first one, then another and another joined in. A slow pulsing chorus mixed of deep male voices and rich female sound, Gregorian in timbre in support of the rich resonant penetrating tones from the flute. As the sounds of the gathering grew, the night came swirling about, gathering its own life force, blending with a slow warm wind from the west that blew at the east and the promise of sunrise, a breath sweet with hope.

Time eluded Greg. He knew he was welcome. What he saw seemed as destiny: aged faces over woven robes, Indian faces, the faces of the Hopi Indians, Keepers of the Promise, last Speakers of the Ancient Covenant—and the female spirit he had seen from the desert floor, Mahrianne.

Amongst the watching, Mahrianne rose to her feet, and as she rose her lithe form swayed with the warm wind, billows of red hair moving gently around her—Greg had been called. His life would be forever altered. As she moved, the dance wove itself around her, enveloping her form, and she became the dance, purveyor of progress.

Grandfather breathed the spirit of the mesa deep into his soul. For a long time he sat blowing his spirit into the flute. The notes floating from the instrument quietly but powerfully told the story of the Five Worlds.

At length the Grandfather placed the flute across his lap and faced Greg. They sat, tired, each feeling his personal thoughts and questions. These two men only now recognized the depth of the challenge. Grandfather spoke first. His words poured out, a stream of consciousness. The words came too fast, one swell following another without a break—the sound of the words, huge, small, sad, and joyous, music of a different kind. After a while he paused, and with a smile of pain he said, "We can only feel what is here within us, feelings of what is here that we cannot see. We feel longing, distance, weariness as we hold the gourd of promise once more. This time, it is you and I who bear the spirit message."

Greg sat cross-legged, rapt in the spell being woven by White Feather. The conversation had begun quietly between the two men, the Indian facing the white man, connecting through their minds. Greg's long-time studies in the paranormal had led him to accept as fact that there was more than meets the eye in this time and space than was generally accepted. He understood the Indian view of life as a continuum where time and space exist concurrently. This day the two men were sharing their wisdom.

White Feather kept the silence of the mesa. His deep respect for all things must be honored. The two men soon began a dialogue in deep perception. The old man spoke first, meshing words and mind-to-mind contact. "Indian Star Walkers transcend conventional bounds of time and space. What allows us to move freely through other dimensions in space and time is the art of shifting our own perception. We must first allow our self to adopt the multidimensional perception. All time and space exist concurrently. You now know this through your new science that demonstrates this as possible. We accept it as basic. There are entryways, or doorways into other dimensions of time and space. By means of these entryways we can go anywhere and bring back what is in that space-time to our space-time perception. Time is no longer linear. It is experienced simultaneously."

White Feather stopped speaking. Greg thought over what he had just heard, trying the holographic view of space and time against what he had learned of recent proofs in his own world of science. He thought he understood. "Then the point is to see the description of whatever or wherever you are and not assign the primacy of one description over another." White Feather's smile and gentle swat to Greg's shoulder was reward enough. "I would like to try it."

"We shall. We will go before you leave Black Mesa this time."

X – DESERT SPIRITS

"He will plant the seed, Thiesha."
"He must convince his friend, but I fear that will not be easy."
"The seed takes time to grow. First it must be nourished. As it grows we shall watch over the young ones."

"Welcome back, Greg." Kate took his coat and closed the door on the crisp evening air. "Your trip to Arizona must have worn you out. A dinner with friends will do you good."

"Thanks, Kate. Sorry Sara couldn't make it tonight. She's taking care of some neighbor kids."

By eleven o'clock the three had finished a relaxed and flawless dinner during which they brought him up to date on Jason's work and the children's activities. When they retired to the soft consuming living room couch and chairs, Kate magically produced three snifters of Cordon Rouge, Napoleon cognac, welcomed warmly by all. Reluctant to push Greg for his story they sat for a while, quiet, content, lazily looking into the fire Jason had laid.

At last Greg said, "I want to tell you about my trip. I know it'll sound preposterous, but it's real."

Kate looked at Greg and smiled. Earlier he'd told them how he felt pulled by some force to explore the Black Mesa, and months ago they'd talked about the on-going genetic alteration of part of the human genome. Now she sensed that while Greg didn't carry altered DNA himself, he'd nonetheless been brought into the Hopi trust. For what reason, she was unsure. She was anxious to hear his tale.

"I went to Arizona specifically to take a look at the three Hopi mesas on the southern rim of Black Mesa—all part of the huge Colorado Plateau. I was just going to look around, maybe see if I could get up to the top of one of the mesas by climbing one of the sides. I just wanted to see for myself what the oldest continuously occupied settlement in the United States looked like. It didn't occur to me that popping in on them like that would be an intrusion until I stood there in the desert staring at it for a while. I know there are better ways to get to the top but I never got that far. I think the Hopi have a sixth sense—I felt the full strength of their powers!"

"What do you mean?" asked Jason.

"I was drawn to Third Mesa, the home of the Fire Clan. Strange, but I don't know exactly how I got to the top. Each of the three mesas is around six hundred feet high. I felt captured by a force of some kind. I could swear I was lifted up to the top by a very old man they called Grandfather." Jason shifted around uneasily trying to find a comfortable position. "Hear me out, Jason. These things happen. Good Lord, you've known me long enough to know that!"

Jason stopped fidgeting. His research had shown that the genetically altered humans possessed different and more expanded cognizance, but he was still uncomfortable with that difference and often felt a little dumb around his own children. Both of his kids had more significant changes in their DNA than Kate did.

"I suspect that what the Hopi shared with me is indeed what'll be coming to the world," said Greg. "We sat on the ground in a circle. The old man, the Grandfather, told the ancient stories of the Hopis that were passed down orally without change through the generations. They now believe that the *Old Ones*, the grandfathers and grandmothers, are the last generation before all that's foretold comes to pass."

"All native peoples have similar stories, or myths if you prefer, that are passed down the same way," said Kate. "The Mayan calendar stops at what would be our year 2012 although they don't know why. Actually it stops at the winter solstice which would be December 21 of *this* year!"

"True, and get this, they all say time stops. I don't think that means there's *nothing* after time stops. I think it means that things will be very different, and possibly that time has a different significance. By the way," added Greg, "our current scientific discipline of particle physics says there's no such thing as time or space but that they're simply some agreement on where the particles of what we *observe* are. The particles of any given object we see are constantly whizzing around. This is what we interpret as time or space."

"In our everyday awareness we include time as an abstract," added Jason, "a river flowing in only one direction. We think of it as the future, or 'next.'

Most aboriginal languages, including American Indian, don't contain a word for *time*, nor do they appear to have any concept of time. Our concept of the passage of time to them is not movement from past to future, but passage from a subjective state into an expression of an objective state, somewhat as if we could move from a dream state to some anticipated reality."

"Another difference, and a very significant one," said Greg, "is between how we view space and how they view it." Greg had studied the Hopis for years and was fascinated by their culture. "To them, space is consciousness; it's divided into tangible, like the conscious mind, and intangible, like the invisible space between tangible things, things we perceive in the dream state. Thus consciousness has two parts and is viewed as a continuum. Interestingly, this view covers the space of the universe as well as the empty space within an atom. Visible and invisible exist concurrently. Night follows day, etc. This is very like the belief structure of most ancient peoples.

"We could probably understand it if we would try to understand and respect their spiritual practices and if we would try to perceive time and space as they experience it. The American Indian spirit world exists concurrently with what we perceive as the physical world. To them, space is similar to the conscious mind and the invisible 'space' between the perceived objects of this everyday world. The unconscious mind permeates all levels of existence. We can't *see* it but we *know* that invisible space fills everything from the atom up through the universe."

"That's similar to the Australian aborigines who view consciousness and space as one," said Kate. "In Australia, a wind instrument, the *didgeridoo*, is central to the aboriginal culture. Didgeridoos are part of their myth of creation. They were used by the original man and woman to conjure up the other creatures, birds and beasts they would need. They're still used in sacred ceremonies, including initiation rites of passage, healings, and other ceremonies."

"There've always been Native American Star Walkers," said Greg. "These are people who can shift their own perception into a multidimensional perception. It allows them to transcend the conventional bounds of space and time by using 'Star Gates' as doorways into other dimensions of space and time. Shifting perception is very difficult for people who've grown up in societies that live by the divisions of space and time. Space and time concepts take a huge effort to maintain in daily life."

Kate replenished their cognac and brought out a plate of chocolates. "I've heard the Hopi stories of the Five Worlds. They're long and intricate with very specific characters and responsibilities for each. We have time so go ahead, Greg, tell us what you learned."

"There were thirteen or fourteen Hopi Elders in the circle," he began. "Each told a part of the Five Worlds story exactly as it had been told before. I felt as one with the great bond they had with each other. It's hard to adequately describe the weight pressing down on the people of the circle. If what I felt as a member of the circle represented what they each carried through time, passed down through the generations, then they represent something deeper than what appears on the surface.

"Off to the side began the sounds of a wind flute. Softly, just in the background, the notes floated from the instrument as the stories told of the beginnings and endings of the first three worlds. Each tale took what seemed like a long time and impressed me deeply. Then the story of the evil of the Fourth World, our present world, and its destruction was told. I wept silently as did the others. A sacred covenant was made with the Hopi Indian Oraibi Fire Clan. They are to be the guardians of the opportunity to make one last try in the foretold Fifth World and the promised new race of humankind, a rebirth of life in peace. That's where we are now."

Jason sat staring into the embers of the dying fire. They had talked a long time this night, these three. The friendship between the two men had grown since he and Kate had purchased the house next to Greg and his family. There were some rough edges but few differences in perception of what bounds there are between reality and the paranormal. For the main part, they had developed a close relationship. He prized having another man to talk with even though he found some of Greg's ideas a bit past his own boundary of reality. Neither of the men possessed the extraordinary abilities of Kate who could sense where conversations were going, a very accurate ESP.

Throughout the evening Kate had listened, enraptured by Greg's account of his two days on Black Mesa. She wished it were time to share what she knew of the mental web connecting the altered people to each other and to the goal of global cooperation and social order. She came back from this train of thought as Jason asked a question about Greg's conversation with the old Hopi medicine man.

"The Hopis regard themselves as the first people of America. Hopi villages sit at the tips of the three mesas that rise sharply above the desert floor. The inhabitants of the Oraibi village on the Third Mesa have lived there ever since people first came to America. The Third Mesa lies to the west, with Second Mesa and First Mesa a few miles to the east. Believe me, it's some of the most inhospitable terrain I've ever seen. They have to hike down a precipitous six hundred feet of steps carved into the stone cliff just to get water. And back up, of course."

"Probably the women get to do that—my guess!" Kate added.

"Probably. There's no water on the mesa unless they collect rain which is very sparse in that desert, although there are springs at the bottom of the mesa."

"And you say they don't doubt the signs that they see?"

"With what they see of the present state of the world and from their charter, the signs are all there," answered Greg. "They're religious in their practice of passing down their messages and they've done that through a thousand generations. The covenant with their god is sacred. Their charter is to pass down the Life Plan and try to influence the people of Earth to live by the Plan given to them by God. The Five Worlds story warns of the consequences of not following the Life Plan for the Fourth World, our world of today. The warning says basically that we screwed up the first three worlds and got slammed for it, and we are screwing up the Fourth trying to get it right. It looks like we have only this one chance to even get a Fifth World in which to succeed! So the final opportunity for salvation is here. This will be the last chance. There will be no more if this chance is blown. God help us!"

"The Fourth World is the one in which we now live," said Jason. "Since they claim that God said the Fifth World is to be the final world, we could easily say that the message is to clean up our act and learn to live together according to the Life Plan. This is also the Christian bible's one thousand years of peace after a great conflagration, isn't it?"

Greg was closer to the teachings of the Hopi. "Well, it's interesting. It looks to me as though nothing is improving. If you believe all things animate and inanimate have a place in the cosmos, then within the boundaries of our lives life must be harmonious, regulated and under the control of minds free of evil thoughts: nature in balance. They believe that a great threat to human survival has been prophesied and that a final war will determine mankind's fate." There was a long pause before anyone spoke.

Jason asked the question on all their minds. "Do we think we're there now? Sometimes it's hard to tell the good guys from the bad guys."

The fire crackled quietly, a peaceful background to their troubled thoughts. Jason sipped the last of his cognac and cleared his throat. "You know, Kate mentioned the Mayans. They had a similar story in their history: *'The waters parted and the tribes crossed over them on stepping stones placed in a row over ground that was all sand.'* This is equivalent to the literal interpretation of the Hopi story that they came to America from the west crossing the sea on boats or rafts from one steppingstone island to the next."

"Another Hopi story is that they came from the ground," said Kate. "Some Hopis believe they came from bottom of the Grand Canyon."

"Well, another theory says that the Hopi might have migrated from Asia to the Americas via the Bering Straight," said Greg. "However, when the Creator gave them one more chance to have a world, He admonished them not to use that back door route, as they would do so without his consent. The implications of this are that if they did, he would again take away their world."

"What do you believe?"

"I believe the account. It's compatible with all the stories and worldwide myths handed down through cultures all over the world."

Quietly Jason asked, "What do you think will happen?"

Greg drew a long breath, his normally high spirits subdued by the conversation. "I don't know. I feel Earth's power and I look at all the torment and changes our planet has been through. If you listen carefully to the accounts of the Second and Third Worlds and their demise, they're all about the same, given that they come from different parts of the world—sort of different perspectives of the singular events. I have trouble figuring out exactly where we are regarding a *demise* of the world or of us, or whether it can be headed off, from the threats man alone has introduced—nuclear holocausts, bio-terrorism, total disregard for the consequences even by the perpetrators themselves. I truly believe we're on the edge right now. We should be focusing on Earth events we know are brewing even as we speak."

Nobody had to ask for refills. Kate was already pouring a goodnight round.

"How would one go about transmitting the story of the demise of the Fourth World as has already been done three times—through the stories that become myths of the survivors?" mused Jason. "It looks like any survivor would have a hard time. Basically start over. And how would we know or even decide what to save from this Fourth World to take into the coming Fifth World? Not likely you could put it on a Blackberry. The Hopi stories and ceremonies have been passed down through time to be executed exactly. What's the story for us? There's no doubt in my mind that we're almost there!"

XI — CATALYST

Every Saturday since she was five, Jessica had accompanied Jason to his office and lab. She played on his computer when he wasn't using it, made a private little hideaway under the desk when he went to the lab, and looked at the colorful charts on his wall. Over the years she grew to know each of the many nooks and crannies in the old desk. She drew comfort from the feel and smell of the wood and its strong presence surrounding her as she played beneath it. There she could let her mind soar and spread its wings. In its solitude she became fast friends with other children, children who lived in places she had never seen.

Throughout the years her favorite toy had been the strange flat piece of stone she found in a forgotten desk compartment. She left it in its container, but often used her crayons and pencils to sketch its strange little stick figure with a tall pole in his hand. Through the ethereal friendships she had made, she had long ago filled in the missing parts of the picture. These days the sketch hung on her wall at home, lost among a mountain of other sketches pinned there as well.

Now 15, she still came to the office most Saturdays. She no longer played beneath the desk, but instead helped Jason with his presentations and papers. As she grew and observed, the data on Jason's walls and the odd figures on the stone formed a pattern in her mind.

"You've been studying the aberrant DNA now since the kids were little," chided Greg. "You've published several well-acclaimed papers and you've given who-knows-how-many speeches, but you've never come right out and talked about what you think is coming," chided Greg. "You're the world authority, for heaven's sake."

Jason glanced at Greg then turned away. "What if I'm wrong, Greg? What then? What about Kate and Jessica? My own family could be hurt by the reaction."

"Kate's a strong woman. Jessica's now fifteen and practically a young woman in her own right. Since when have you shied away from criticism? Other scientists have questioned your findings in the past, but you've always been able to convince them that your research is sound. Look Jason, in the decade in which you've been doing your DNA research, I've continued my research on Earth changes. Everything I've seen convinces me that forces are accelerating. I know the myths and the stories of the world's religions, and everything falls in line. I believe the old stories are about to come true."

"Maybe Dad was right," said Jason. His shoulders slumped as he knit his fingers in his lap.

"About what?"

"He thought my science was snake oil. I should have gone into teaching like my brother. Then Dad would have been proud."

"Your dad's gone now Jason, and besides, you know better! He may not have understood what you were doing, but he'd be proud of you now, believe me. Now, about this speech you're going to give at the annual genetics convention . . ."

Jason grinned in spite of himself and prepared to counter Greg's arguments all the way.

Greg began the debate, as their conversations often became, between his survival-oriented view and Jason's scientific mind. "You say almost eighty percent of our DNA is not mapped to anything we can relate to our characteristics. It's considered *junk*. This discovery was a surprise. Some refer to genes by characteristics, like the *God Gene* or the *Immortality Gene*. Science shows that an observer affects a science experiment just by observing it with his particular prejudice toward the outcome. This is what quantum physics is all about. A person creates his or her own *reality* simply by incorporating consciousness into the equation."

Jason interrupted before Greg could draw another breath. It was his speech after all. "Hold on Greg. You're getting carried away." Greg ignored Jason's bid for control of the conversation, his own thoughts off in another direction. So Jason sat quietly, smiling a wry smile at his friend. "Go on," he acquiesced.

"Okay." Greg mused a moment. "This, then, is a mind that 'creates.' The part that's both frightening and reassuring is that a mind's acceptance of a prediction will most certainly assure it will be true in its lifetime. Time virtually collapses in on itself, and the body is a rendition of the mind at a distance. This is one of the great conundrums of physics—creating a temporal asynchrony, a singularity, seen as rifts in time and space."

"Are you saying that a person can predict the future or get a glimpse into the past?"

"Exactly! There are documented experiences of personal items lost in the past apparently physically *falling* back into the present, or of clear visions of the future presented to someone."

Jason preferred to work from hard data, yet didn't want to occlude information just because he wasn't familiar with it or held it in question. "I don't want to debate this. There's not enough hard data in existence yet to avoid argument. Let's just leave it that some people now and through history momentarily see future events or even become snared in a time not their own."

"You mean like Jesus?"

Jason thought for a moment. "Who knows?"

"I issue you a challenge. You know more about what's going on in the current genetic melee than anyone. What you know has been published. It's now open to debate. You're the only one who can bring sanity against the fear people now suffer. What's coming out of you is yours. Somewhere, as the result of running through the corridors of your mind, opening doors and slamming them shut, searching for a meaning to your life's work, you opened one and were blinded by the light of the truth as you perceive it. And you knew this was it. Now, are you dedicated enough to handle it? Are you dedicated enough to present to the public what you believe, and are you tough enough to take the criticism of those who don't understand and yet not become cynical? I hope so!"

"Okay, okay! I'll write the speech but I haven't yet decided to give it." Jason stood and shook hands with Greg, partly to show that he wasn't irritated and partly to move on to his next task—writing his speech. He also wanted his assistant to make a list of people who had direct involvement in key aspects of the project. "I'll call you as soon as I have something. See you at dinner tonight, about seven. I hope our ladies will have something great prepared. They said it was a surprise!"

As soon as Greg left, Jason called his trusty lab assistant. "Linda? Can you come in here?" *There's definitely a pronounced differentiation in the distribution of enhanced DNA.* The gauntlet now lay squarely at his feet. Jason knew his friend was right—there was only one choice.

No Linda. He picked up the phone and called the lab but again there was no answer. Oh, yes, Saturday, her day off. No wonder.

He knew that genetic discrimination could threaten individual privacy but that was just another problem he'd have to sort out over time. Right now what he really needed to do was to get started on a disclosure release from North Western.

XII — GRAVITY WAVE

The cold light from the terrified face of the man in the Moon shown brightly through the windows of the darkened plane, illuminating the unsuspecting faces of the sleeping people.

Kate noticed the change, probably a slight pressure difference in the cabin of the 747. Nothing to worry about. She glanced around at the other people, some sleeping, some reading. A few stirred slightly in their sleep, adjusting positions then resuming their steady breathing pattern. Someone snored. A few heads turned or looked up and then went back to what they were doing. Yet as she looked around the faint light that had filled the cabin softly changed to lavender tinged with green, then turned rosy. No, it must be her. She closed her eyes. Maybe she needed some sleep. The weekend symposium on children's diseases plus the plane ride had been tiring.

She couldn't have been asleep long, probably just dropped off for a few minutes. Yet again, something seemed different. Kate opened her eyes. The rose-colored light in the cabin had shifted to a soft golden glow, comforting yet disconcerting. The woman in front of her reached up and turned off her reading light. Kate pulled down her shade. Same change. Quickly she pushed up the shutter on her window and leaned forward to peer out into a familiar bright starry night. What met her eyes was unexpected. Most people had become accustomed to the luminous arc of golden light that now circled Earth night and day. The light had come from the north, circling itself like a tendril around the globe and rejoining itself at its point of entry. It appeared to retain the arc from whence it came and keep its closed ring as Earth turned under it, no apparent explanation for its presence or for its tether to the center of the universe.

But now, if she looked intently, a dim but faintly glowing band of light spread out in both directions from the arc: a shimmering iridescent structure of flickering lights in more colors than Kate could ever have imagined. She sat stunned, transfixed on what appeared to be a living growing structure. Some part of the light ring could be seen day and night somewhere in its passage through the heavens.

The plane pushed through the sodden air laboriously making its turn into the airport in a silvery arc, too slow for a plane, she thought. At what speed a plane could no longer sustain its weight?

At that same moment Jason felt it, a ripple, just a slight ripple, the forewarning of the force behind it, a wave, coming across time, oblivious to space. Something alien coming, sweeping across the galaxy, emanating from the core of the universe—irrevocable, collecting, analyzing, the sum of all through which it swept, gaining strength and knowledge.

The Universal Mind was expanding to fill all known space and time, reaching beyond its own strength, becoming the sum of all it encompassed, moving beyond any bound, gathering all information, all wisdom, all feeling, all goodness, all wickedness.

It accelerated as it expanded, heading toward oblivion or implosion and enveloping the man standing at the railing of his deck. Jason stepped back from the railing, knuckles white where he had gripped the rough wood, his heart pounding in his chest, throat taught.

He knew that the mathematics of our universe was required to support a broader, universal math that describes the determinant aspects of our universe. But he wondered whether ours could be the only universe, or could there be an infinite number of universes related by yet another set of universal laws? Perhaps life is but a hologram sustained by a perpetual wave . . .

It was ever so slight, just a small sensation of increased weight, just like what she had felt a week before on her way back from the symposium. Kate stopped loading the dishwasher and lifted one corner of the translucent flowered kitchen curtains. A small wave of some kind of energy, *a gravity wave* Greg had said. Jessica and David were standing together staring north, or at least Jason thought it was north. Neither Kate nor the children moved. The three of them were stony still and unaware of their surroundings.

"What the hell is it, Kate? What are you three looking at?"

Kate didn't reply, something Jason had come to fear more and more lately. He leapt to his feet overturning his chair and scattering the peas he had been forcing from pods. He rested his hand on her shoulder and the two peered out the window. Jason thought he'd heard a sound but couldn't catch hold of it. He'd felt something pressing down as the wave passed. This was the third time in less than a month this had occurred. Physicists he knew were happily embroiled in quantum physics and arguing whether these waves were the forerunners they had envisioned of a space-time made up of sets of *branes* defining our universe. He questioned whether anything less than two Planck widths could exist. From somewhere deep in the pit of his stomach fear crept through his body and into his soul.

Kate turned and mustered a small smile. "It's okay. It's not time yet." Jason knew better than to push it so he began retrieving the peas he had spilled. He felt, at some deep primitive level, that all humankind was being slung to the sidelines of a great cosmic dance. The instinct of all humankind welled within him, and he knew. The unbridled mind of creation, the intellect that created the Sun, the planets, all matter and antimatter in the universe had begun an unstoppable expansion. A burst of final, infinite creation had begun, rushing to a balance of matter and antimatter, the standing wave of an infinite void: that which had formed the universe and created Man was creating again!

XIII — THE DREAM

"He's wavering, Thiesha!"

"Watch and wait, OmRa. Perhaps he will see the way. She has already seen the future."

For the twentieth time, Jason reviewed the notes he had made for his upcoming speech. What if the notable scientists and thinkers he invited to the symposium laughed him off the stage? There'd been many naysayers over the years who had been happy to criticize his findings in journal articles as well as on TV interviews.

Hell, I know I'm right, thought Jason, but is it worth the grief I'll stir up? I could lose my funding. If that happens, all of my work for the past decade could go down the tubes. It's just not worth it! He tossed his notes on the desk and rubbed his burning eyes.

Kate pushed open the door to his office and stuck her head in. "Am I interrupting anything?"

"No. Come on in. I was just working on this speech Greg keeps pushing me on. I think I'll just stuff it. I'm tired of the whole idea."

"Don't do that quite yet." Kate sat down and pulled her chair up close to Jason. "I'd like to share something that's been bothering me. It's about a dream I had last night. Do you feel like listening?"

"A dream? Sure. Anything to get my mind off of this mess."

"The dream was so vivid. It was more like a premonition. I know that sounds overly dramatic, but it's been on my mind all day. It was about the ski trip Sara and I've been planning."

"That's probably why you dreamed about it."

"True. It's probably also connected to all of the things Greg's been sharing with us about the Mammoth Lakes area he's been tracking, not to mention the hot changes I've been making on my car." They both laughed, knowing Kate had beefed up her BMW to a race-driver's level, but her smile quickly faded.

"I dreamed that Sara and I drove up to Mammoth along Interstate 395. The drive was quiet but something gripped me in my soul. I pulled off the main highway and onto this desolate stretch of road wondering where it ended and why anyone would want to live like this.

"We followed the road up a small hill that stopped at what appeared to be the opening of a mineshaft. As we stood there thinking about the old gold rush miners, we noticed movement in the earth, a small wave, just one small undulation in the vast expanse of the high desert. An earthquake? Where? How big? Surely it was just a slight movement in a desert area known for its frequent small adjustments. Besides, we were still quite a distance from Mammoth—no worry there.

"We ran to the car and sat for a moment before we started laughing. Compared to one run down Mammoth Mountain, a small earth shudder is nothing! So I turned the car around and stomped on the accelerator to make the sand and gravel spew from under my big new radials. We left S-shaped skid marks on 395 as my Michelins dug in.

"There was plenty of snow when we got to Mammoth but things seemed different. On the first day there we both noticed the constant but barely perceptible feeling that the snow-covered ground was moving, undulating in slow faint waves. We asked around for some kind of validation or denial but the lodge music continued and everyone appeared happy. The next morning we took an early morning run before starting down to load the car.

"I remember sitting quietly, looking at the mountain with my mind. Before long I could see a pattern of symmetrical ripples around the mountain and the subsidence of the ground where the ripples were forming. Sara saw it, too. 'I know,' was all she said."

Something flickered across Jason's mind then disappeared. For a moment he was confused. He had to force himself to focus in again on her story.

"Greg had told us that the calderas have been inactive for 760,000 years but a bubble of magma builds up over time and blows out at regular intervals. In the case of the Mammoth calderas that act was a hundred thousand years overdue. Predictability on a human time scale is ludicrous, but with the rising mile-deep, half-mile-wide basalt bubble, Sara and I decided it was time to go.

"There was no time for a phone call. I was going to hook up directly with the kids and let them know what was going on and that we were on our way home, fast! I knew how frightened they would be. In my dream I knew they were safe spending the weekend next door with Greg and Robbie. I also knew Greg would tie in to my thoughts and fear. I hoped he could connect with you as well. The kids would see everything anyway, unfortunately."

She paused for a moment, gathering her courage to recant the fearsome dream. She felt compelled to tell Jason about it. Why, she wasn't sure.

"Sara told me she felt a presence, a pull. She was scared! She said our daughter was warning us and that we had to get out of there NOW! In my mind I saw Jessica. She was screaming, 'Mom, Mom! Run! Go fast now!' She was crying. My God, Jason! It was a nightmare! We ran to the car leaving everything as we ran.

"Our run to the car and the drive down to 395 seemed to take forever. I yelled to Sara to get in, close her eyes and not to touch the wheel. I knew Mammoth was going to blow. We had to be at least fifty miles away before it did!

"Sara rode beside me staring into space. Jason, it was eerie yet calm. It was space without time. It was connected to that unseen power you can touch only when your mind lets go, when you hand over all feeling, all sense of control, all erstwhile knowledge. Her hand was on my shoulder, light and steady. She was our conduit between two spaces—no more complicated for her than finding the children as they skip among the higher dimensions of space and time."

She no longer looked at him. Her voice had changed to a soft, rhythmic chant. Jason stared at her, his heart loud and fast in his chest. He felt like calling her to come back from wherever she was going but nothing came out.

"Time stopped," she continued. "I joined the unseen path she formed and closed my eyes, tuning into dimensions beyond humankind's palpable four. I could see the children as I screamed for them. They heard me and stopped dead to face the beast pursuing us as we drove.

"A spirit wind hustled down the valley blowing between our car and a great hot gray dragon. It was laughing and snorting, rushing to consume its prey. I felt the strong hands of you and the boys fold firmly over mine melding them into one machine. I heard you scream, 'Look straight! Look straight! I'll steer. Hang on!' I felt your heart pound, your stomach cramp.

"We were blown by the wind of the devil, blown by a fear so deep that time stood still. I pressed hard on the accelerator—90, 100, 125, more, more, 150. My brain processed so fast that the roadside seemed motionless. The black pavement with its guiding white spine glistened, gluing my Michelins to it, sucking them hard with its ownership. Self-preservation heightened our awareness changing the dull mauve of the desert hills into brilliant colors. Faded green plants alongside became bright green-gray objects. The grip on my soul was guided by a higher power, the strength of a giant holding my hands to the wheel, a child's game of road race in a penny arcade.

"I glanced in the rear view mirror. That's what you're supposed to do, isn't it?" She turned her head slightly as if to ask. "Every eight seconds glance into the rear view mirror to see what may be bearing down on you? I started counting; one, two, three . . . the long drive would get tiresome. Home would be good, Jason, Jessica and David. I glanced at the mirror again, creepy now; the skin on the back of my neck recoiled as my hair brushed against it. I shook it off of my skin and watched the sky, its gray mass moving and changing in unnatural and eerie ways. It was nothing I'd seen before. I wasn't keen on seeing it now.

"The elaborate movie playing in my rear view mirror was surreal. Time slowed. The landscape vanished. My world turned into a slow-motion drama, our lives the pawn. The roar of the demon chasing us invaded my mind. I tightened my grip on the steering wheel. I was rigid, a cutout in this bizarre chase.

"Fortunately 395 is straight as an arrow in its march south along the Sierras. We knew when the blast of air driven by the monster behind us would hit. Our car was being pushed along. Our lives depended on my keeping control of the car. The hot gas cloud from the eruption would come behind the rushing air of the pressure wave. We knew the volcano probably was the first act of a longer play. It was only the introduction to the violence of the calderas lying under Mammoth Lakes and Mammoth Mountain. My arms, wrists, elbows and shoulders throbbed from the effort and added pressure.

"Eight! In ultra slow motion I looked again into the mirror. Then, only gray. My thoughts spinning, time slowed even more. I hit the accelerator again hard, hoping there was more speed left at its command. The gray hit, pushing us forward, faster—to where? Only gray in front now, and gray had a face. Against my will I closed my eyes and held on to the steering wheel gripping it with all my strength. The awareness of other hands holding mine never wavered. I felt heat, hot searing heat. I heard screaming, and I went blindly forward screaming into the enormous gray face. My black BMW disappeared into the dusty mass—two red eyes, questioning."

Kate's voice trailed off. She slumped, drained by the effort of recollection.

Dream or no dream, Sara's mind bridge had worked better than she would ever know. Jason usually couldn't remember dreams, but Kate's story had brought back a forgotten memory that now came forward in blazing detail.

But seeing the way Kate's dream had affected her, Jason had no intention of sharing his own with her, at least not now. "Look, honey, it's just a dream," he tried to console. "Or rather a nightmare!" he added. "Lately we've been focusing too much on these things. Let's go home and see how our teenagers are doing."

For the next few days Kate's dream lingered in the back of Jason's mind. His own dream now gnawed at his thoughts, leaving him curiously unfocused and unsettled. Was it the same night as Kate's nightmare? He remembered sitting bolt upright in bed, roused but not awake from an exhausted sleep by something odd.

A dream? It was vivid. He'd actually believed he was there in it. He'd been standing rigid on the desert floor facing Black Mesa towering high above the desert. He knew he'd never been there, yet the howling wind that greeted his arrival stopped abruptly. His only sensation was the violent awareness that Kate was in trouble. A billboard of thoughts scrolled quickly before his closed eyes. The car. That damned car! His gut told him she would test that performance soon, but not by choice or for sport.

Jason clenched his fists and tried to replay what he'd been feeling. He was standing on the soil of a spirit desert. His left foot braced against something, it didn't matter what. His right foot pressed the flat of a narrow gas pedal into the sand. He was driving that car, seeing what she saw, feeling the manse of the car, hearing the whine of its powerful V8 engine and the scream of the road.

He could hear her favorite tape, Savoy Brown's *Hellbound Train,* bridging from its opening rhythm into its long pounding beat synchronized with his wife's heart. He felt and heard the violent rupture of a tortured Mother Earth whose children had forsaken her needs, broken her trust, and that now, rent in agony, spewed forth molten rock and scalding gasses as far as she could. The lava bubble hidden by Mammoth Lakes could no longer be contained and Kate was its prey.

High on Black Mesa the old Indians felt the call from the spirit mesa of Jason's mind. They moved to the edge of the mesa softly chanting, softly tapping their drums, drumsticks padded with animal skins. This steady simple rhythm slowly decelerated as time moved forward, their paced and firm drums slowing time enough to let the angry Earth's prey escape. They remained thus, focused on Jason and the black car somewhere in the western desert.

Jason felt the Indians around his fallen form dripping cool water into his mouth, the wind soft in the afternoon sun. They whispered to him the story of the Spirit and helped him to see, adding an invitation to come back in the future. The visitation had melted away leaving Jason sobbing on the side of the bed.

All the while, the Owens Dry Lake presented its dull pastel pallet to travelers. Its once milky blue waters had been harnessed to wet the tongues of Los Angeles' industrial growth of the last two centuries. On Highway 395, the booming town of Lone Pine rests at the gateway to Mount Whitney, towering over Owens Valley to almost fifteen thousand feet. Serene in its beauty, it remains geologically active. Just north of Lone Pine a monument marks a memorial to the casualties of the great California quake of March 16, 1872. At 2:30 a.m. the quake struck leveling the town and snuffing twenty-six lives, most of the town's population. It was a magnitude seven point nine.

The golden granite boulders of the ancient Alabama Hills rise from the desert floor of Owens Valley. Chiseled for centuries by wind and rain, they have been sculpted into a massive spirit domain that speaks to a lone visitor come to hear the stones. In the deep silence of a hot desert day, the ancient act of laying-on-of-the-hands connects with the ancient ones trapped inside the stone. Slowly the magic dance begins and they begin to speak. The one who has come to commune feels shrouded in dense warmth, no sound is heard save for an occasional warning rattle from a slithery passerby, or a bird questioning, watching the dance. Visions, those who once dwelled here, rattles and drums, a wind flute, deep heavy heart-sore sobs—all gone, all changed; the weight of things to come, oppressive, heavy, mortal—passing from the rocks into the human. No one can walk away. It takes time. Eons of time.

At night the cool clean air, ruffled by the patterns of heat waves rising from the rocks, glittering, laughing, knowing, freeing the collected day's heat, shares the bright panoply of the heavens as far as the eye can see.

The blur of the ancient weathered rocks, watchers of Earth's slow changes, obscured their wisdom. These tellers of secrets to those wise enough to listen, these Alabama Hills, lying alongside of the youthful Sierra Nevada Mountains, would once more be witness to the performance of underlying pent-up violence, this time triggered by the fickle performances of the volcanic calderas surrounding Mammoth Lakes, a famous ski area on the east face of the lofty Sierras. Mammoth Mountain is the nervous remnant of a volcano that last blew up 760,000 years ago leaving enormous calderas. The mountain lies at the western side of these calderas. Its spotty outward serenity belies the ability of the volcano to destroy once again.

In the nights after Kate had shared her terrible dream Jason held her as if each were his last. If she felt his desperation, she did not show it. Serene and calm she gave her love completely.

There was no longer any question. Under a cloud of impending loss, Jason worked furiously on his speech. Was he losing the one woman he had ever loved? Could he be losing his beautiful daughter as well? He was scared and desperate. There was no way he would lose them without a fight. If he had to lose the people he loved, maybe he could help others in the process. The message of his charts and graphs was undeniable, his thesis sharp and clear. He would tell the world what he had found, let come what may.

XIV — CHOICE

Jason's uncertainty evaporated as he mounted the stairs to the stage of the great amphitheater. He placed his notes on the podium and confidently faced the audience, now more certain than he had ever been that this was the time. He stood quietly for a moment under the canopy of stars and breathed in the cool, clean air of the hills ringing the seats. In the first few rows he saw the faces of those he loved most as well as those of his trusted colleagues and friends. He cleared his throat, glanced once at Greg who gave him a thumbs-up, and began.

"Colleagues, ladies and gentlemen, you have been invited to attend this meeting as representatives of a broad spectrum. You are all highly qualified in your areas of science, ethics and religion. Please hold questions on issues until my talk is finished. I assure you we will have ample time for discussion. Tonight I wish to share with you what I have learned and now believe to be defensible." Jason put aside his notes, moved to the front of the lectern and spoke directly to the audience.

"You have all become accustomed to the Light Ring encircling Earth." A few faces turned toward the heavens acknowledging the glowing ring. "About four billion years ago, the First Light, or the beginning of time as we know it, wrote a message, a chemical message, which became the continuous thread of all living things. That message is still the secret of life. Somewhere along the way, the fundamental mechanisms of reproduction on this planet were worked out, and hence we are here now.

"What we know today is that something has touched the process of evolution, probably for the entire human race. We know that something is accelerating changes in the DNA of children coming into this world. Apparently it has been occurring at some level for at least two millennia.

Something is *waking up* dormant DNA, altering or augmenting it, or some trigger is activating a pre-coded program in our DNA. Genetically, we are becoming a new species of humans."

Jason turned towards the giant screen behind him and switched on the presentation he had prepared. There were murmurings of surprise, shock and disbelief as well as a few loud scoffs. Quietly he scanned the faces in the crowd waiting for the murmurings to subside. "If you will bear with me, I will explain my premise.

"The indigenous peoples of today, ostensibly the survivors of the last destruction of the world by water, teach that there is no real life or real death. The totality of life experience here on Earth is that of a dream within a dream. This would be an analogy of *living* within a hologram sustained from some source of which we cannot know. Today's mathematics and physics come closer and closer to a theory of the structure of the universe that supports these ideas. No matter how you choose to look at it, whatever set in motion our universe and sustains it is beyond our ability as dwellers within the hologram to fully comprehend.

"Some believe an intervention has been going on for several thousand years to help us survive the enormous threat facing our planet. I believe that now. I did not before but the evidence is too great. The occurrences today fall into step with predictions noted in Hopi Indian lore as well as predictions of others. We can no longer ignore these warnings. The *Great Purification* is here.

"The work of deciphering the human genome marches on. The probability that all of its intricacies will someday be known is growing. Changes in the molecules of DNA occur for many reasons. Some of this we now understand, some we do not. We are making truly vast steps forward in understanding the mechanisms and probabilities of various changes. The thought has occurred to many people, particularly those involved as keepers of our souls—clerics, philosophers, people who cannot get past laying total responsibility for the state of humankind on a god—that we play with the proverbial fire as we pursue our boundless quest to answer the question, 'What are we and why are we here?'

"It may be possible that the changes we are now seeing in the psychic capabilities that some possess, capabilities that would be far beyond those of our recent ancestors, are due to changes within DNA wrought by some different mechanism than we have knowledge of now. That appears to be true. These changes in the DNA sequencing of some people cannot be explained by what we observe today. Even powers of healing or supernatural knowledge now appear in almost all populations.

"In addition, we have the Ring of Light forming around the globe. You have to ask yourself what is happening and why. If you look back through the most important works humankind has ever produced, you see evidence of rare changes: the prophets of almost all religious works document existence of unique people throughout time. Personally? I think we are being given an opportunity for some great step forward in our capabilities as a species."

As Jason spoke these words his voice began to blend with a greater voice, soft at first, then growing as his words poured out over the gathered people and transmitted around the globe. "Just maybe, all of the changes we see in our children and ourselves have been sped up because some other intelligent life form is reaching out to us. Perhaps we have now reached a place where we are worth saving."

The audience, quiet now, sat in awed silence, these men and women not only of God but of science, some with eyes closed, others peering far into space, seeking, communing with an innate belief far stronger than myths of men, in an origin and a destiny with whatever form their gods took. For they knew, at a cellular level, that they, all things flesh and otherwise, were but variations on one theme, be it some cosmic vibration, some sound, profound, beyond mortal belief, or too intimate to be accessible to tinkering. Whatever the cause, this starry night would be the end of doubt. For humanity, mass alteration of human DNA had continued to occur, albeit slowly over generations.

Jason's words had blended with the soft but powerful voice now increasing in resonance. As it assumed the roll of speaker, Jason raised his face to the heavens and fell under its spell. He saw the Light Ring increase in the depth and beauty and gave himself to it. The voice poured richness and warmth over the assembled. No one doubted that it was speaking to the entire world and with words appropriate to each difference in voice or meaning.

"The Light Ring growing around your Earth is an inter-dimensional door. It provides access to synchronization with the heartbeat of the Galaxy."

OmRa's voice, resonant and warm, spoke to each soul of Earth individually, rich with truth and terrifying in the expanse of the message being delivered.

"Each of you will have a choice to make. This choice will determine your participation in the affectation of changes necessary to your introduction as Citizens of the Light. As your year 2012 approaches, many will learn to live 'in-the-moment,' to live taking change as normal and expected. Changes that worry others will be taken for granted by these steady people. They will be the ones preparing to manage whatever is needed to maintain life under new or stressful circumstances brought on by the expected changes on your planet. The notion that time and space are illusions created by a human need for order will be their flag. These people will remain through the events that will reshape your Earth and nurture her through the changes. Not all of them will survive Earth's changes."

A pause of softness, a feeling of tenderness poured subtly over the listeners. Succor and love warmed the waiting. A small space in time waited for these words and their message to be fully absorbed. Then, with a barely noticeable shift in timber, the voice of Theisha chimed in softly prefaced by a shower of sound from tiny trilling bells. The voice breathed cool fresh air scented lightly by some soothing fragrance. Onlookers felt a hand gently caress their cheek, a loving gesture, reassuring. The words rang musically, a melody to be followed by deep, deep introspection into each soul on Earth.

"Many of you have been waiting, souls altered generation after generation, for this event, for this choice. You have changed spiritually in many ways and have believed without total conviction that one day, veneration of beliefs would occur. It has. Not quite as you had believed, but closer to truth. You have walked the passages of the soul's deep knowing and quiet. Others have looked at you wondering what it is that you know. What is it that guides your life so differently? Why, you wonder, do they not see what you see? You who follow this choice will rise to the Light Ring to be prepared for a different world when you return."

Soft calming sounds filled the quiet space. At measured time Theisha continued.

"This is your time for a new world for people choosing to evolve, a time for joining the wave of evolution with the force of the future. You long for a connection more than simply amongst yourselves—a global connection is closer. However, we reach one step further. We reach to the stars, to the brilliance of our Galaxy of Light. Your choice is between remaining here on your Earth or joining the Ring of Light at your choice of time. The Ring of Light will remain where it is for the necessary time. At the end of that time you may either return to Earth with what you then know or move on to the World of Light. You might think of it as a rite of passage, a shift of consciousness."

OmRa presented the third choice.

"The third selection from which you may choose is simply to carry on as you are now. No changes. No choices to be made. Your life will not change from the path on which it is set.

"If you do not elect to choose from these three, then the third choice will become your choice. The call to choose will be heard round the Earth. The choice will be read from your mind. The only choice to be implemented in phases is the choice to go to the Light Ring."

So there it was. Three choices, and choices must be made. OmRa had changed the freedom to choose into a mandate.

Ecstasy and mystery engulfed the crowd, and a great *om* rose up. It filled the space between Earth and sky reverberating, bouncing back from the Light Ring now circling the globe.

The shimmering figure appeared simultaneously around the Earth in the place between sundown and sunrise. She stood in the light, moving slowly, placing her feet as she turned, hips moving in slow undulating circles, head thrown back, arms extended skyward, palms down, fists clinched, head now moving slowly down, face smooth and drawn back, eyes tightly shut, unaware of any but the pulsing music absorbed in an ecstasy of united mind and body.

Jason and his colleagues had delivered genetic proof of a new race of people. Now, the audience waited in personal silence for the mystery of the Light Ring to be revealed. Which of the three choices would each choose, and how would they make their choice? Some pondered in silence the meaning of a call home, even knowing they may surrender the womb of Mother Earth forever when they heard the call, knowing they would forsake those who could not or would not hearken to the call, maybe forever.

The women in the crowd smiled quietly. Yes, the women smiled. The women of the world had always known. They said nothing, simply understood something they had always known. And their children danced free in the midst of chaos.

XV – Loss

Neither could eat their food at dinner. The kids didn't seem to notice although Jessica was quieter than usual. That night they listened to the children talk about their high school friends and activities before heading for bed. Then they sat looking at the lights in the valley below, holding each other tightly, trying to absorb the very essence of the other. Jason was too sad to notice the mild excitement in her voice or the joy in the small but constant squeezes she gave him. They did not speak much, nor could either, absorbed in their own fears and hopes. But as the evening passed both fell into an intense awareness of each other, a total absorption in the other.

"We will meet again, you know," Kate whispered. "Maybe not the same, but we will know each other."

"Yeah, I guess," was Jason's only reply. He could barely collect those few words into some semblance of a sentence. Unwillingly, both arose from the couch. Jason dimmed the light and arm in arm they climbed the stairs to their room, knowing they would love each other one more time, one more time for perhaps all eternity.

As Jason bent over to remove his shoes a warm tear fell. Kate could not contain her soft sobs as they lay together. She cradled in his familiar arms pulling her closer. He wanted to etch her into his body as she was etched into his soul. Through an unbearable sadness, a deep feeling of loss, he could sense her joy wrapped in tears of separation. His confusion profound, he was overwhelmed by his need for her, to possess her as his own one more time, and very softly, barely a wisp, his warm man hands consumed her communicating what a man never says.

And through her anticipation of ascension her soul responded to his need. Enmeshed in the bliss of soul mates, two separate entities joined as one, now moving forward, trusting in faith to spheres unknown and unknowable. They lay there the remainder of the night, Jason relieved by sleep, Kate exhausted by exhilaration.

She sat up carefully to keep from waking Jason, but then he slept soundly, the sleep of the innocent, she mused. She laughed softly. *Yes, he sleeps soundly.* In the seventeen years of their marriage he had never wakened to her night wanderings or for much else. She wondered what would wake him before he was ready.

Kate gently swung her legs over the edge of the bed and watched a familiar faint indigo blue light cast itself over the room and blanket all she could see. The light spilled over the balcony and softly coated the sweeping view of the hillside. Their second floor bedroom let the soft night air flow in. She rose and stood by the bed, once more looking at Jason, wishing he could travel with her. The deep indigo cast pervaded everything, penetrating, wandering.

When the pre-dawn display painted the ceiling with peach, she walked to the window, opening it wide to let in the brightening colors of light and the crisp cold reality of the fall air. *It is time,* she thought. *Winter will be here soon and new realms are calling. Mother Earth's time is near and there is much to do.*

She knew her feet would have reached the floor but there was no sensation that they were touching anything. Neither was she clothed in her pink nightgown. She stood by the window clutching her folded arms. A sad smile passed her lips playing with the corners of her mouth. Hot tears had somehow snuck in and were stinging her eyes. *I know I put on the pink one,* she thought, *but I don't have it on now. I have something on but I can't exactly feel it and—weird! I can't seem to see it when I look!*

Jason opened his eyes sleepily, momentarily unable to comprehend what he saw. She smiled at him as she passed the bed gliding toward the door of their room. In his half-awake mind she seemed somehow different, a little airier he thought, an illusion of the light probably. His mind rejected what his eyes told him—she wasn't walking, she was drifting. Her body was clothed in a filmy garment, transparent, billowing and flowing. He was sure she was fading from his view. He couldn't speak, frozen in his anguish yet happy for her passing to the Light.

Jason leapt from the bed tangling his feet in the covers as he attempted to follow her. He stumbled free from the bedding and ran toward the door. Only then did he realize the door through which she had passed was closed. He yanked open the door and yelled her name. No answer. He raced the few feet to Jessica's room only to find it empty, too. "Kate! Jessica!" he called, choking and sobbing.

David bounded from his room calling, "Dad!" He put his arm on his dad's shoulder and gave it a loving shake. "Dad, they're gone!" Then, quietly, consoling his grieving father, "Don't you know? It was their turn. They'll be back. Mom said so."

He stumbled out onto the deck and into the cool night air. *Get used to it. They're not here.* Again Jason's thoughts ran wild, out of the grasp of reason. *There is a force! There is a power!* He looked down, hot tears dropping on brown knuckles crushing the life out of the wooden railing.

His logical mind still couldn't make order of the situation, and for the thousandth time his mind stepped into a loop trying to grasp reality. First, and basically acceptable, was the telepathic ability of his wife and the kids. Then, the one, but odd, thing he had become accustomed to was their kinetic ability, their uncanny knack—no, that's not the right word—their apparent parapsychological ability to be in several places at once. Their ability to make contact with other children, each in some other part of the world, this world at any rate, unnerved him.

The parts of this rumination he feared the most and understood the least were the circumstances around their unions that had resulted in the conception of their two exceptional children. Now David's resolute determination to remain with him rather than meld into the Light Rings with his sister and mother had surprised and comforted him.

He watched the clouds gather sheet after sheet, glowing fingers of dawn spreading their grasp, growing cauliflower clouds, white as cotton. Dancing sparks of light silhouetted against their darkness. A backdrop for their ethereal beauty, each small bead of light chose color from the Ring of Light or from the rainbow emblazoned over the heavy dark clouds. Jason wondered if he would be able to distinguish Kate and Jessica when their turn came. Probably not. Perhaps they would buzz him showing some acknowledgement of connectedness. He joined the thousands who now spent their spare time contemplating the powerful display in the heavens. With the scent of her floating in the air he tried not to scream for her to come back. David moved quietly to his side. Had he noticed the wisp of White Shoulders that hovered about, waiting?

"How will they find us, Dad?" David echoed his own thoughts and concerns but knew there was no answer. "Here, Dad, I wrote this note."

Jason looked at his son's words and fought to maintain his composure. What he wanted to do was pound on something and scream! Knowing the words on the paper would push him over his limit, he said, "Please, David, read it to me."

Quietly, painfully, David read. "Dear Lord, please save our world. We have no place else to go. We did not mean to hurt it, but we did. Please."

Jason bit his tongue. He took the sheet of paper gently and read it again to himself.

"Dad, when are they coming back?"

"I hope soon, David. I hope soon. Most likely when the changes are over and Earth has settled down again."

Alone on his deck that night, Jason again smelled the perfume she had always worn. The silence of the night deafened his brain. The only sound was that of distant thunder, audible and reassuring. He felt his hands tighten their grip on the rough wooden railing. Hold on, he muttered to knees that were melting, abdicating their task of supporting his body. He sank to the deck still gripping the rail, rested his forehead on the wood and sobbed, great wracking sobs shuddering the length of body, pulling at his arms. The knot in his belly grew tighter, waves of nausea swept over him. Involuntarily his body convulsed wrenching his numb hands from the railing and he screamed her name, over and over.

He saw Greg standing quietly across what now seemed a massive chasm between the two houses, watching, weeping, screaming softly to himself, unable yet to grasp his own loss.

That he could not be the familiar comfortable master of himself Jason knew. Sobs won over screams and he lay there knotted in doubt and fear. How had he ever thought he had a grip on what was happening? How had he ever *consented* to let her go? But it had never been his decision. And the kids! *God! Please, what have I done? Help me, please. Help me!*

A soft hand smoothed his brow and quieted his gut. It stroked his hair, caressed his face. Soft arms cradled and soothed him. He opened his eyes expecting to see her bright face and silky hair next to his. No one was there. He closed his eyes again. No one was there!

XVI – Four Tablets

"Time," Thiesha mused, connected to OmRa, mind to mind as they were. "Think in Earth years, act in Earth years. We have only so much time to complete our task. The web layers are in place and the foundations are solid. The alterations remain in the gene pool, but the carriers of the early layers are mostly gone now. After the transition we will have to work on that, but we must complete our work. The changes are in motion. There is only a small time left in Earth terms."

"Time?" OmRa mused. "We cannot interfere."

"What we have done is an intervention."

"Intervention is just planned interference."

"The tablets are coming to Black Mesa, OmRa. They will be quietly gathered and brought together. As the tablets of each of the four races come together, the remaining worldwide changes will be set in motion. The new Life Plan will be drawn for all humankind. They have earned our help in keeping mayhem to a minimum, but we cannot ameliorate all of it. There will be massive changes in the world. Before that time, there is still much to be done."

"For two thousand years we have seeded the main backbones of humankind with genetic change. Mahrianne must now call these youths to the Mesa."

"She has done her job well. Let us watch."

Absently Jason fingered the old desk. Alone in his office he was comforted by its solid warmth. It had been his anchor to reality in the days since Kate and Jessica had gone. Even David hadn't been able to help lessen his grief. He'd tried, but without success, so he slowly pulled away to grieve by himself.

Jason poked at the bottom drawers with his foot. What had his gruff old father said about the contents of that odd box, some sort of ancient artifact? Jason thought about the stone and the strange drawings. They hadn't meant anything, only part of a broken whole. Curious now, he bent down to pull on the drawer and found it ajar. He smiled remembering how Jessica had played with the stone as a child and how she'd sketched what she thought was the rest of the pictogram. He peered into the darkness. The stone tablet was nowhere to be found.

Sometime before the winter solstice young Wind Runner had noticed a thin strand of gold in the night sky. At first he wrote if off as an aberration, maybe the reflection of a hair. Grandfather watched, too. He smiled. "Soon, grandson, soon."

Though the calendar dates did not match, it was near the solstice when the other voices began to join their conversation. High in the Himalayan Mountains a youth looked to the sky. "Do you see the gold thread?" he asked of a young woman drying her flaxen hair in the sunshine of the Alps. Mind to mind they spoke as they always had.

"Of course," she replied. "We are not alone in our observation."

"That is true." Jessica added her voice. "I feel a call. Do you feel it, too?"

Scientists at the Perth observatory in Australia thought that the faint gas trail in the night sky might have been left by some particle, but they were busy with more important things.

The haunting notes of the wind flutes floated across Black Mesa and out over the vast expanse of desert, a pleading cry to the spirit world. Soft pulsing rhythms pushed the accompanying sounds of muted drums of longing and waiting to the edge of the mesa and nudged the notes into a flow of welcoming sound. Notes that carried qualms of anticipation of some ancient event passed down across the ages carrying the message of deliverance and duty.

The legend was yet to be fulfilled—it prophesied that preservation of the old ways would be the only means to survival. Its message told of the solemn duty to live one way, to preserve what the ancient gods had given as a sacred trust, to be executed at some unspecified moment in the future. A tireless trek through life after life, soul after soul, with only the faithful guides—owl, hawk, eagle, wolf, bear, and buffalo—always there, always encouraging, keeping faith through the centuries, carrying yet a great sadness with their joy and sacred trust.

A week had passed since old White Feather had begun to notice the accelerating messages about events leading to the winter solstice. Mother Earth had fared as expected throughout the rapid changes. She was now at a crossroad of evolution more poignant and severe than at any event recorded by humankind.

Grandfather's mind remained connected with Mahrianne's as she prepared the three tablet bearers who would be joining Wind Runner, the fourth and final tablet bearer. Together they focused their minds and constructed a path of silver light that stretched out from the mesa. It branched into three paths, luminous beacons for the travelers to follow on their way to Black Mesa.

The weight of the task at hand remained carried by the Tribal Elders, and their hearts rejoiced—so many years, so many tears, no more fear. Now, at last, their sacred covenant would be fulfilled. They prepared for the ritual ceremony that would bring about the uniting of the foreseen Life Plans for each race and for the worldwide unification of the four races.

One by one the three chosen youths came to Black Mesa transported swiftly across the living desert and escorted by Mahrianne and her Light Workers. Young Wind Runner greeted each as they appeared and made them welcome. Through the collective power of the sacred knowledge handed down through generations and held by the Tribal Elders, the connection among the four tablet bearers would now become complete. Now close friends, as their contact with each other through their growing years had engendered, they rejoiced at their first physical presence.

They were soon introduced to the Elders who immediately took over the management of the four tablet bearers. Housed together under great secrecy, the youths were given time to comprehend their task and to compare their tablets before the ceremony of completion would be held.

White Feather waited and observed, sharing with them the joy and enthusiasm of their embrace. They were now aware of the heavy expectations laid on their young souls and were determined to play their role as directed. The Elders waited to give them time, but not long. There was much to do as the role they would play was for all humankind—the bringing together of the Fifth World inscription on the four pieces of the final tablet. No one knew for certain what tasks might spring from that union. But the forces of centuries of waiting would be unleashed. Not all would be gentle as corn silk.

Serene in the fulfillment of her task Mahrianne watched, savoring the unexpected thrill of joy and excitement. This was their time now, the four youths she had helped into this world and guided through the long preparation for this event. These youths represented an awesome responsibility. Her four! Her sacred trust to all humankind. The joining of the four tablets would begin the final cataclysm as the worldwide family of ocean tectonic plates and spreading zones readjusted to their new positions. These tablets were a message to all of humankind via the mediator—the Hopi Fire Clan. *Choose the higher path to purification so that your ignorance will not ultimately destroy your planet.*

Moon Shadow had stopped to smile at her son and the joy she saw as preparation for the ceremony of the tablets became a whirl of activity. The haunting sounds, plaintive yet whole, spilled over the mesa providing a soothing connection between the activity and the beat of her heart. In her heart she knew this spell was being woven for her son as the collector of the tablet pieces and matchmaker of the four tablets. She knew the prophecy—it told that the required life story for each of the four races was about to be fulfilled.

The Elders slowly formed a circle around the four youths who sat in the center facing one another in a circle of peace. Stories told that the four tablet bearers would share their teachings at the end of the Fourth Cycle and tell of their contributions to the Life Plan of the Fifth World.

Others watched the Elders and the youths, chanting to the soft sounds of the wind flutes as they rocked side to side. The taut skins on drumheads decorated brightly with bird feathers and painted skins provided a low, slow ceremonial beat for the dance of the Elders. One by one the youths would place their stone fragments on the earth accompanied at each step by the ritual dancing and chanting of the Elders.

The first stone was laid by the black hands of the youth at the circle's western point. Her stone formed the beginning of an ancient pictograph. The Old Ones' eyes closed as they moved slowly one step, then two short steps, and again, one slow step, then two short steps, over and over around the small circle. They would dance like this through the violence of the extant Earth changes, casting a safe shell for Black Mesa and her fulfillment of duty.

From the northern point of the circle, the white hands of the alpine youth reached out to set her stone. The pictograph began to emerge. Moments passed. Almost imperceptibly the drums deepened their beat. At precisely one turn around the circle the dancers reversed direction and repeated the movement.

Slowly, from the southern point of the circle, the Tibetan youth held out his amber hands and placed his stone with the other two. Drumbeats grew stronger. Dancers quickened their steps.

Lovingly, reverently, hands the color of beautiful red earth raised the fourth and final tablet above the three waiting stones. Then gently, quietly, he placed his stone with the rest completing the pictograph of the Life Plan that so many centuries ago had been scattered across the world. At that moment a small shudder shook Black Mesa, unnoticed by all except the Tribal Elders.

The image now complete, the young ones smiled, elated and relieved that they had fulfilled their role in the changes to come. The sound of the drums intensified, now louder and stronger. It filled their young bodies with the great spirits of the desert, reverberating across the canyons, echoing from the smaller mesas, covering Black Mesa with its shield.

The dance of the Elders would continue unabated long into the night. They did not tire nor did their age interfere as they repeated the moves and rhythms that had been passed down through time. They had only the retold stories to guide them in their movements, stories unaltered through an ancient past, legend and myth passed generation to generation.

Now, without bidding or words, the Tribal Elders had begun the final dance, the dance of the coming together, the dance of duty through what may have been centuries of time and pain. They would guide these young people through the preparation and the years necessary for the once-only opportunity to pull humankind back from the abyss of the destruction accumulated by self-righteousness and fear. They would not fail. Their chanting and dance would last three nights and days, a catharsis, summoning the forces waiting to initiate the new world that would arise from the ashes of humankind's selfish childhood. Childhood has ended. It was time to join the Galactic World of Light.

XVII — Waiting

The twinge that now gripped his chest reminded him that today was December twentieth in the Year of Our Lord, 2012. In Jason's mind the image *12/21/2012* appeared. He struggled to force the icy hand gripping his heart to let go and turned his eyes toward the heavens, away from the bowl of earthly gems. And as his eyes adjusted to the night sky, the luminescent band of light now surrounding the globe of the Earth glowed softly, its iridescent hues forming patterns of exquisite beauty, shaping and re-shaping themselves as the Light Beings prepared for the coming awakening.

The malaise had been building in Jason, slowly at first, then accelerating over recent months. It now turned to fear, fear that was shared throughout the Los Angeles area. Earlier today there had been a brief announcement from the Cal Tech Earthquake Center. A magnitude 2.6 jolt had been detected near the dogleg meeting of the great San Andreas Fault and the Pacific Plate. These two giants were locked in mortal combat, the Pacific Plate moving counter clockwise toward the north and the Continental Plate moving inexorably clockwise to the east. At that critical point they locked together and fought to break free from the other's grip. The growing pressure would yield to the tenets of physics, slipping along the fault line they created.

Jason stood on the point of his deck and peered toward what lay beyond Los Angeles to the east. Though he couldn't see it through the haze, he could feel its presence. Two weeks ago he had begun to feel the pressure in his head building. Until that time he'd been too busy for the power of the moment to catch him. He ran down his checklist, ticking off completed items, and realized he was ready, well, ready as anyone could be. Panic edged its way into his brain, his breathing accelerated.

As the minutes passed, tightness gripped his body crunching his guts. Beads of sweat now dotted his forehead yet he shivered in the chill of the moonlit night. Rivulets of sweat wet his shirt and clung to his back. In a futile attempt to choke down the fear he raked his hand across his damp forehead.

It had been a clear cold day in the City of Angels, one that only Los Angeles seems to have, one that breaks bright and dazzling after the late fall rains, so bright that without sunglasses the effect is like snow blindness. Jason stepped out of his clothes, naked in the growing darkness, enjoying the primal feeling slowly invading his mind and body. He had worked hard to remodel the house riding high on a hillcrest in the Santa Monica Mountains, a place above the noise and smog, an aerie for his children. He dipped into the steaming hot tub and let the water flow over his cool skin in a wave of warmth and welcome. The hot tub on their second-story deck had been his wife's idea, one to which he had fortunately yielded.

He looked down over the basin. Streaks of cotton fog lay over Playa del Rey and parts of Santa Monica obscuring the light on the Palos Verdes Peninsula and framing the glitter of the lights in the city below. Between him and the bowl of lights sparkling below, black silhouettes, cut-out trees and houses formed a wall of presence turning the city lights into an enormous bowl of precious gems. Red, green, gold, and bright white blinked and shimmered as waves of the captured day's sun escaped into the chill air.

For months the light band had been growing wider and thicker as more beings joined it daily. Jason wondered how many so far. Silently he called the names of his daughter and his beloved wife and their minds answered quietly. *Watch, father, they are coming now.* Peering intently, he could see a new golden ribbon of lights, not exactly light, more like brightness comprised of millions of showers of smaller lights, billions perhaps. From this great distance it would be impossible to tell. *See? They gather. The universe accepts us into the Light World, our new World of Light. We shall return soon, you will see.*

Jason relaxed then into the warmth of the water feeling a slight tingling sensation as a new band of light cut across the luminous belt now around Earth. It was as if a ring of power had emanated from the heart of the universe and passed through Earth's light band hooking it at a right angle. This outer arc of an energy ring radiated from a central point such that its essence passed through the light band and then held firm to it. Thus two rings hooked together, one passing round Earth, the other, infinite in length, crossing through it in acknowledgement of ascension, a hook of connection and control.

He heard a tone, first one, then joined by legions, a sound not heard by man before, felt more than heard and indescribable by any terms or analogies Jason could conger. It filled him with joy and fear. Rapture descended over him and through a space-time vortex that allowed him to see the future, he saw tomorrow: twelve, twenty-one, twenty-twelve— Earth time.

During the night the quiet came. Jason stood in the darkness listening as the church bells chimed twelve times. At the twelfth tone he looked up at the radiant band over his head. His heart full, he smiled and blew a kiss to his loved ones. Softly he whispered, "Easy, young Mother, we who wait will tend you." The enormity of the thought struck hard. She sat immobile in time and space, wandering in a collage of infinite universes, and the Moon hid its face behind the shadow of the Earth.

A short wave radiobroadcast to no one in particular droned on, skipping and stuttering among the structures of man. Just one more pitiful pre-planned efforts of technology, Jason thought. Quietly, he sat on his deck looking down on the twinkling lights of the Los Angeles basin.

The voice droned on. "Killer tornados ripped across the US Midwest today taking out concrete and brick buildings in the towns of Pierce City, Missouri, and Jackson, Tennessee. Damage was widespread over a large area of the Midwest striking with winds exceeding two hundred miles per hour. So far, forty-two people are known dead and scores more are still unaccounted for."

The faceless weatherman continued. "Tornadoes of magnitude four to five are rare even though the Midwest is known to have up to twelve hundred tornadoes a year across 'tornado alley,' a wide area in the path of the tornadoes. Tornadoes spawned in this stretch of territory spin to alarming speeds as they blow northward where they strike havoc year after year. Besides leveling concrete or brick buildings they often make matchsticks of homes and businesses in their path. They are known for leaving total destruction in their wake."

That's not all, thought Jason. His mind wandered to something Greg had told him. Hurricanes were known to form off of the African Coast near the equator and follow prevailing winds on Earth anywhere from the east coast of South America north of Brazil, up to Canada.

The two men waited, each alone yet oddly connected. A faint glow tinged the deep night sky above, neither interfering with the stars nor obscuring the lights of the city's basin below. Across the canyon they exchanged a quiet glance. Both Greg and Jason felt the excitement of a challenge welling within themselves and an odd bond of mutual need and trepidation about events to come. They had worked together over the years to establish a solid support system for their city. Now they would be called upon to make it work.

The night air breathed the faint odor of ozone, of ionized air, a soft brush of electricity against their skins as the glow in the night sky deepened and brightened, now blue-green, iridescent, then softly swirling. A distinct hum coming from everywhere, voices of a thousand million humans, resonant, thrilling, diffuse, surrounding, penetrated their flesh, their souls, their very essence.

Threads of iridescent blue-green descended from the swirling mists above, coalesced from the shroud of electrified blue-green light enveloping them. Each sought connection with the Light Forces, for to remain through the anguish of a tortured and twisted Earth, these builders of the new world would give life and hope to others whose task it would be to build again upon this new genetic step a new race of humans connected with their brethren throughout a universe, bound in one consciousness, one mind, one love.

A thread of light reached Jason. He felt it brush his head, center top at the cowlick. As it touched his head he was filled with loving energy, peaceful and safe. He looked at his friend now kneeling and clutching the railing of his deck. Greg was experiencing the same—the thread of blue-green light now entering his friend's head also divided and threaded over his body as a stream of water spreads in a fan as it passes over a stone shattering it in a shimmering veil of reflective spray.

The gray came in the wee hours of morning shielded from view by the very darkness of the new day. By the time most folks were stirring a dead calm was rapidly contributing to the illusion of impending suffocation. Sound seemed dull and dead, oppressive. Radio commentators searched for some explanation for this strangeness. "It started in Kansas right on the flat ground, no warning, none at all," said the man. "It felt something like a stomp from a real big foot. I'm covered with dirt and blood and I'm wet through and through."

Jason stood feeling the massive pulse of the beast sprawled below him. Reflexively he touched his wrist. His own pulse coincided eerily with that of the great metropolis throbbing audibly below, alive, vital, and massive. An undertone of fear began to well within him beckoned by the odd but unmistakable coordination of his pulse with the monstrous throb of the city. A perpetual orgasm of machinery fueled by millions of motors and wires that, obeying the laws of earthly physics eerily flowed in wave patterns like the synchronous ticking of random clock works in a clock shop. Intuitively, not in sync but left undisturbed, speaking only with the rhythmic disturbances of the clocks, inanimate, they vied for supremacy—massive drums of energy from another world, another time and space.

Jason's heart raced. No pounding like this in his chest had ever occurred no matter how far or fast he had run. Somewhere between panic and awe, adrenaline had flooded his body and washed over his mind.

The voice he heard sounded far away, calling his name, "Jason! Jason!!"

He watched Greg from his deck knowing that his help was neither needed nor wanted. He had been there himself and could only watch tears streaming down his cheeks, the depth of his friend's agony reaching deep into his soul, his very essence. They, these two men, keepers of the young Mother, would be thrown together for a long, long time.

Mother Earth began to rock herself gently, rocking, singing a soft faint lullaby. Tears flowed hot from sad and frightened eyes over her entire surface, except for those still aware, still awaiting their shuttle to the Light. The young Mother hurts. Cannot you feel her pain? Cannot you share her spasms, the stretching pains of stress?

Jason stepped back from the railing, knuckles white where he had gripped the rough wood, heart pounding in his chest, throat taught. He fought for air. Suddenly it came, gasping, panting, sweat soaking his shirt, more than the chill of the clear night should have supported. And he knew, in one fell rush, at some deep primitive level he knew. The instinct of all humankind welled within him and he knew—the unbridled mind of creation, the intellect that created the Sun, the planets, all matter and antimatter in the universe, had begun an unstoppable expansion. Entities in space, keepers of the future of humankind, the mind that had created man, the power behind these strange and exciting changes . . .

A burst of final, infinite creation had begun, rushing to a balance of matter and antimatter, the standing wave of an infinite void—and Jason was thus forever committed to his role in the great new dance. Unknown to him or anyone else, the beginnings of vast change had been here a long, long time.

For one fleeting moment the world below suddenly quieted. Jason touched his ear thinking he had been deafened by the sounds of destruction. A squirrel poked its head from under a fallen power pole and squeezed his way free. After shaking his fur and testing each limb, he made his way up the lengths of power cables now crazily tilted. He remained there, front paws on one cable, hind paws on another. He didn't light up and fall to the ground as Jason expected, toasted by the current. Rather, his primitive brain had registered the total power failure. A tired smile crept reluctantly across Jason's sweaty face.

XVIII — SHE AWAKENS

In a remote South American jungle a playful pale brown monkey sat in the sun grooming herself atop a stone structure rising above the surrounding tangle of trees and vines. She was startled by the sound of an abrupt crack, a rupture in the stone upon which she sat. The heavy circular stone had been placed vertically near the crumbling top of an ornately carved structure. Archaeologists interpreted it as a calendar attributed to a vanished Mayan culture. Mysteriously the dates on the calendar stopped abruptly at a specific year, a year that translated into the current calendar as the year 2012, twelfth month, the winter solstice—*December 21, 2012.*

A wave of bright yellow light brushed the creature now clamoring wildly, unsure of her perch and wanting to continue preening in the sun. The monkey screeched in irritation at being distracted from her grooming. Taking a tight hold on the branch to which she had jumped, she allowed the rhythmic rocking of the great tree to lull her to sleep. Tomorrow would be another day . . .

Deep in the heart of Haleakala, the House of the Sun, Pele stirred, her long sleep over. Just a slight murmur, very slight, and then she quieted. A brief report was logged. Nothing more.

As the Sun God sent his fiery orb into the heavens casting his first rays deep into the Haleakala Crater, Pele stirred once more. The rays of the Sun touched the silent silverado plants inhabiting the spoon-shaped bowl of the crater, and Pele stretched again. The stretch split the floor of the crater and left a crack one hundred feet long.

Rangers, responsible for guarding the safety of visitors, performed their jobs quietly and efficiently evacuating everyone except those geologists authorized to record her events. The geologists, men and women of Earth, unknowing in myth, made notes and sent messages. These people of Hawaii understood and respected Pele and quietly performed their duties as scientists. Then, silently, they began their long descent into the volcano and its mistress. Their helpers followed quietly, breathing a soft goodbye to Pele. There was much to do and time became uncertain. Pele stirred again and the men and women hurried. Roused from her long sleep, angry at the inconvenience, Pele shook off the comfort and concealment afforded her by her mantle.

Where were they? she asked. She saw no sacrifice, no baskets of fruit, and no payment for such inconvenience. Where are my gifts? She rumbled and the ground moved again. Her ire grew. No bright feathers? No golden ornaments? No spears or warriors to pay her homage? She stretched, a bit out of sorts after so song a sleep. Still, no one noticed. And her ire grew becoming a molten fury deep in her fiery heart.

The park rangers froze, unsure what had happened. Then, in the gray of pre-dawn, they saw smoke rising from inside her crater, the mother seat of a violent and unpredictable Goddess, Pele, owner of Haleakala, House of the Sun. Pele was getting restless, uncomfortable with the fire deep in her belly. Angered by the lack of respect paid her, she commanded the great crack on Hawaii's southwest face to let slide its entire mass into the Pacific Ocean. A tsunami generated by this enormous mass was propelled across the Pacific, unseen until it reached shallow water where its feet tripped and stumbled on the sea bottom. The first wave formed and then began to propagate freely. By the time it reached the Americas it would spread from Los Angeles in the north to Panama in the south. In the center lay Puerto Vallarta—damage would depend on the wave's power upon arrival.

Pele, still angry and outraged over the lack of respect shown her, abruptly turned to face the formidable Mauna Kea that by now was receiving gifts and prayers from the frightened inhabitants remaining on the island. But, alas, there was no maiden as a sacrifice to her whim. Pele spun around twice, spreading lava and brimstone over the island. Then she stomped her foot leaving only rushing, swirling water where once an enormous part of the island had lain. Pele had shaken the world's largest volcano abruptly awake, leaving not a thing standing nor any creature alive on what was left of the island of Hawaii. The remaining inhabitants of the far-flung islands were left to cope with the swirling hole in the sea as the backlash of her wrath.

A girl stood on the eighth floor of an old luxury hotel in Puerto Vallarta, a feather brush in her hand. She had just finished preparing the room for hotel guests. She didn't know what made her stop and peer out the window yet she stood there for some time. Looking down from such heights made her queasy. She felt something might be wrong with her, but she wasn't paid for standing around. She took one last quick look at the scene below. People were just standing or wandering around, then as one they turned toward the sea, expecting what she did not know.

Abruptly it stopped. No more odd feelings. With only a quizzical look she turned and went back to work. The old creaky balcony, though small, didn't give her much room to maneuver. As she moved the few pieces of outdoor furniture, a pressure wave of some sort nudged her forward. She lost her balance grabbing the railing to keep from falling. Another pressure wave passed through the structure of the building moving it just enough to loosen the rusted support pins of the railing. She grabbed what was left of it, then saw something her mind couldn't grasp. What she thought she saw couldn't be. The ocean was changing shape, a living, breathing monster born of the Pacific Deep sent on a mission of destruction across miles of pliant water. The figures below were running every which way gesturing to her about something she couldn't make out.

Something was wrong, something that felt alive, that grabbed her in the pit of her stomach and sent a wave of nausea up to her brain and back down to her already roiling gut. She was not alone. People along the western shores of the Americas looked at the ocean.

Far to the west the Japan Trench yielded to the massive pressures between two moving tectonic plates, the Pacific Ocean Plate and the Continental Plate, as if God's samurai severed that tenuous link to safety in one blow all the way to the Kuril-Kamchatka Trench. Parts of both main islands slid into the abyss of the Pacific Deep. Segments of the Great Trench began pushing both the North and South American Plates.

On the other side of the world the pressure generated by the great African Plate where it pushed against the smaller Arabian Plate was now increasing. In one way or another, the movements of all tectonic plates were responding to pressures directly or indirectly from all other plates. As these interactions reached points where pressures exceeded resistance, earthquakes and volcanoes provided temporary relief, a giant living jigsaw puzzle.

In the deep deafening silence Jason thought his eardrums had burst. Then, slowly, barely audible, a low soul-wrenching sound rose from the chaos below. Stinging tears welled in his eyes as nausea gripped his guts. Everything was moving, the jerking, shattering precursor of a great quake, a settling shivering movement. *Sweet Jesus!* The whisper hissed through his clenched teeth and taut lips.

Then it began, a sound that would continue four days and five nights, a sound that would etch itself on the soul of anyone who might survive this night of the brutal retribution giant. A mortal sob built to a screaming crescendo—it would persist until the chosen would be sorted from those remaining and the healing could begin.

Jason had no words. He wretched over the railing, convulsing and at the same time cursing any god angry enough to visit such anguish on so many. He'd had no idea. Had he known, he might have elected a different path. Privately he was glad his wife and daughter were with the Light. Jason turned to look for his neighbor but saw nothing. Then, ignoring his own fear turned and fought his way into the house and his son.

Unmindful of the carnage transforming her skin, Mother Earth continued heaving and undulating in her ecstasy. Only the deep throb of change within occupied her focus, this softening into convoluted reverberations caused by the chaos of polar shift. On her surface, humankind clung to whatever came their way, hurt, frightened and angry at the Mother's uncaring throes. Above, the Light Ring pulled and caught any soul bearing the slightest mark of awakening and held them close, fixed above, observing the magnificent display below, a difference of perspective and awareness of the opportunity for all of humankind.

A boy and girl stood transfixed by the change around them, unmoving in the glow of the heavenly spectacle, unshaken by the tremors and convulsions under their feet. And the multitudes fell to their knees, hands clasped over ears, bowed in some inner fear, finding no escape. Others stood silently, almost hearing the sound, motionless in their incomprehension. They heard what they did not hear. Though immobilized by it, they were unaware of it.

The Great Mother's cry of primordial pain, cold and hard, squeezed White Feather's heart in synchronization with some deep limbic brain of the planet. The scream began deep within Mother Earth, an outcry from a rebellious Earth unable to prevent its carnage. At first her cry was a long painful moan. Her anguish deepened and her cry rose and pulsed as she convulsed in her agony. Again and again she wailed as rock scraped rock, scratching, crumbling, then trailing off in an unbroken wail—the wail of a woman chained in torment, unable to free herself or save her young. Then, as if in retribution for some heinous crime, she moved and twisted. And bent by unstoppable tearing, Mother Earth screamed.

Black Mesa stood firm, resolute in a broiling sea of red dust, impervious to the screams of the desert below as she complained and adjusted to the chaotic changes in underlying bedrock. Her roots extended to the center of the planet and were not a thorn to be removed. Black Mesa stood, one block of land in space and time, quiet, while Earth beneath began to slip and cry. The time of the Fifth World had begun. Grandfather was now locked in communion with ancient powers and ready for the coming changes. One by one the waves of change would come locked in step with the unstoppable power being unleashed on Mother Earth. Those who had been chosen as enablers would become conduits through which the forces of time imposed the path of change.

The great *om* began softly, a breeze coming from everywhere, going everywhere. "*Ho Ka Hai* my brothers. *Ho Ka Hai* my sisters," White Feather whispered in his native tongue, ancient words showing his people's way of expressing both beauty and commitment. *It is a beautiful day to die!*

Quietly the web descended.

XIX — EMERGENCE

"Your time is now, play your part well. Humankind rests in your soul. Link now."

OmRa's precise message echoed round the world, a thousand, thousand times, a call to those carefully linked together, a conduit through them to the thousands who could hear and feel the rapture of the tones now being radiated to those who would lift humankind to its next level and return it to the new planet, genetically ready to begin anew.

"The link holds, Thiesha. They are strong enough this time, the magnificent children of the blue planet. The collective mind of the universe remains and will be forthcoming. They will need encouragement and assistance for some time. This is not a child's task. We prepare now for the next phase."

The awful silence belied the devastation below. It hung heavily in the pre-dawn air as a shroud stunning Jason. He knelt there on the broken pieces of wood and glass trying to understand why there was no sound. Maybe a blast had ruptured his eardrums, he didn't know. The enormous gap of sound made him feel he was in another world. He'd expected to hear the sound of buildings falling apart and crashing to the ground in twisted masses of steel and glass or pathetic mounds of broken timber and concrete.

He clapped his hands together to see if he could hear them. When he did, his heart jumped to his throat and froze there. He reached for the banister he'd been leaning on and tried to pull himself up. He fell to his knees again as it gave way under his weight.

The oppressive stillness was the most disturbing. Jason first thought he'd been deafened by the brutality of Earth's agony. He rubbed the large bump above his ear. What he could see in the cold dawn light didn't look right. Little shudders continued to disrupt his efforts to stand. Gray light slowly opened the chilly sky over Los Angeles. He peered at the field of orange lights, now on again, beads in a bowl of darkness. From his home, or what was left of it, the lights looked familiar but different. Pieces of the night's events came seeping back into his throbbing head. The bowl of lights tilted as if some giant hand were tipping it toward the Pacific. He stared at the beads, dazed and disoriented.

From where he'd been standing before it began there was only darkness all the way to the orange beads. Slowly pieces began to fit. His home high on Mulholland Drive gave him an expansive view of both Los Angeles and the San Fernando Valley. The orange beads were the west end of the LA basin where it meets the sea in Santa Monica!

"Dad! Dad!" came a frightened plea through his thick silence. *Who? Oh my God!*

"David!" He mouthed the words. "David! Are you okay?"

"Sort of . . . I guess so," came the shaky response.

They flew into each other's arms and clung there desperately in the dim light of a horrendous dawn. From the twisted masses of steel and concrete rose a moan, a maniacal, mechanical moan like great teeth gnashing together. As they knelt there trying to steady themselves, the sound began. Pale at first, rose another sound, a soft sound, an ache that would not be put down, the sound of pain, and the sound of fear and bewilderment. It started softly at first, just a wee faint sigh of a sound. It rose up and out and beyond. It invaded the recesses of the universe. It was the cry of the people of Los Angeles. Jason wished he'd never heard it. He wished he were dead.

Later, exhausted from the long ordeal, Jason and David summoned their tired souls and cast their wet eyes upward. The shimmering iridescent golden rings still shown brightly. The painful rebirth over, the new Mother lay quiet.

There *was* a tomorrow! The transition was utterly unexpected and as gut wrenching as the tortured cry of Carlos Santana's *Black Magic Woman*. Blackness threatened to descend on Jason muffling the wave of nausea invading his body. From the ecstasy of survival and the challenge of what lay ahead, a great crushing sorrow spread through him, through his very soul and brought him to his knees.

While Mother Earth lay sorely wounded, a soft blanket of snow had fallen upon her fevered face, giving her strength to begin anew. The clouds glowed bright with a light of their own, each infinitesimal droplet of moisture a tiny prism to the glory of the Sun's evening rays. Volcanic dust in the air wickedly enhanced the broad spectrum of the last light of the day, emphasizing the reds and oranges as if to claim kingdom over her color display.

The great *om* began softly, a breeze coming from everywhere, going everywhere. Then as more of the altered ones linked into the Web of the World, the sound began to swell. As it swelled its timbre became rich and full, minds and voices from all races melding together in one great wave. The wave resonated and grew, filling the air, penetrating all matter. It was joined by those watching and those waiting, cheering in different worlds, waiting for humankind to take the next step, joining by whatever means they had to boost the frequency of the great growing *om*. And this pressed the huge mass, bearing down on the bright blue and white planet, pressing it into a different path, focusing on the chaotic birthing spasms of her beautiful blue body and white flowing hair, the streaming white hot magma of her heart, preventing it from destroying her life-giving skin.

At the beginning of the tone, silent sets of hundreds gathered, some in small groups, some in larger groups, some alone, each with their own instruments, and they began to play. The sounds they made grew slowly joining in a dance of communion, weaving in and out, now pulsing, now sustained, building and growing into a powerful vortex of sound that provided a focal point for the entry of the ancient promise, fulfillment of an ancient covenant by ancient instruments.

The band of lights began to sparkle, a myriad of tiny lights. Jason could swear they were speaking to him telling him it would be okay, and he knew this was true. As the band sparkled and shimmered it began to rotate, opposite to that of Earth he thought. Faster. Faster. A soft sigh emanated from it, through it, around it. As it gained momentum, the sigh rose to the murmur of voices, now chanting, now singing, now softly cooing, soothing as to a newborn child just retrieved from the ordeal of birth. As the wheel of light settled into a steady state, glorious ribbons of colored light, ever changing, began to fan out from it, rippling in paths reaching to Earth, caressing her softly.

Jason's mind blended with the ribbons of light and he no longer felt the cold of the December night. His thoughts were strange; they invaded his mind, a kaleidoscope of stars, miscellany of images not identified and pictures at an exhibition of the future. Aeolian pipes played to an ancient rite, *Hear, comes the bride!* — a glistening mesh of the Light Rings, a wedding band of light for the new Mother. A Light Ring coiled and twisted, a spiral closed in on itself, a giant double helix bearing evidence of the gifts of change given to humankind.

Across the world in the growing light of day, a shadow crossed the crowd, just a fleeting wisp of darkness. Few looked up. She was one of those. A chill blew through her soul. She turned from the crowd and made her way up the hill to its crest. A few joined her there. They stood in silence, no greeting passing their lips, only a glance that penetrated into the souls of each who joined them.

Again she turned from the crowd and sought refuge, this time beyond the shimmering light. She rested her hand on her breast, feeling calm spread through her, the eye of the hurricane. No one moved. Eyes closed, the small silent group faced the crowd assembled below, seeing through a single blind eye. Magic ruled the air, an oppressive shroud over the assembly. At first no breath stirred the air, no sound escaped their mass. The wind swirled at the feet of the few, small dust devils of surprising violence. They did not notice. Eyes still closed, the arms extended forward, palms up, waiting.

More wind, now picking up small twigs, bits of the dry grass, foaming with the dust of the desert, tugging at loose garments and strands of hair. None moved. Now brooding, now seeking to be noticed, petulant, angry, the wind moved faster, circling the small group standing oblivious to its striving, focused on the unwary assemblage below, searching for any indication of another who did not belong there.

A small child held in anger by its parent struggling to keep it still, quieted and turned its wide eyes back over its parent's shoulder, searching for the source of the growing sounds. It smiled as it heard the sounds floating down from the hill, blown by a rising wind, enticing it to reach away from the restraining arms and hands. The parent, more and more captivated by the entrancement smothering the assemblage, lost focus on the wriggling child. Warm hands, bringing tiny tinkling sounds and sweet tones like a candy flute, reached out to the child. It cooed with delight and wiggled free of parental arms, reaching upward to the sparkling hands.

Received into the warmth of the hands, it was immediately withdrawn from the parent and the doomed assemblage and drawn quickly up into the small group focused on the mass below, waiting. With the child cradled in the arms of one of them, they turned in unison and hastened down the far side of the hill. It was only then that they looked back toward the top of the mount they had just descended.

Those who remained below now saw light in the heavens, wide bands of lights, indefinable in terms they knew. Silently they watched, some sitting, some standing, some even lying on the ground watching the heavens. As the lights slowly appeared, the sound of a breath, sweet and melodic, quiet, barely audible accompanied them. People stopped what they were doing, setting aside the tools of their trades, the aids of their homes, the toys of their wars and came to watch the extraordinary event taking place.

Earth herself was quiet, waiting, facilitating the event. She stopped her rumblings, her spasms, her shaking, her raging of storms, an awesome stillness disseminating calm and stimulating a cautious sense of awareness across her face. Behind all this a single tone, rich and resonant yet barely audible, began to rise. From everywhere at once yet from no discernable focal point, blending with the melody of the breath, a rapture began within the inhabitants of *Die Erde*. The Earth.

As they watched and waited, those in darkness lighted by the heavenly display, and those in the light of day, overwhelmed by the vision in the midst of their day, there formed a band of living light stretching from horizon to horizon. It joined light and darkness where the line of separation between night and day spins, a complete band of light around Earth, a band of enormous proportions and made of the light from the World of Light that unites the universe.

Through this breath of time, the single tone began to swell. The people of Earth were breathing with the tone. Slowly in, slowly out, rejoining the resonant vibratory tone of the heavens. Unknowingly, the people of Earth were breathing out their part of the universal *om*, the *om* of the tone swelling around them, sustaining the Ring of Light around the Mother in her moment of creation.

"The Light Ring is formed. It is time."
"The World of Light awaits you. Welcome."

Their words rang like a bell through the World of Light, faint at first, a cry from the collective soul of man locked in knowing something unknowable. Once again, Aurora's bright mantle painted the sky with bright flames of purple, then red, then peach and finally the golden glow of dawn.

XX — Reborn

"They will find abundant food in the flowers of their gardens. Though they are yet unaware, the foundations for their support have been formed."
"This planet's Fifth World comes now into being. The planet is reborn."

The aura around the child lifted her. She made her way slowly down the aisle of the cavernous room, her shoeless feet leaving prints in the soft pile of the carpet beneath them. For one who would notice, the prints cast their own pressure on the floor, not so much as glowing marks, but as ever so tiny sparkles to a watching eye. She drifted, motionless, along the way to the altar and came to the softly carpeted steps. Her lithe form was wrapped in a garment wafting in small billows, blown gently by a breeze that touched no one else in the great hall.

To the crowd gathered there, waiting that they might see her, came a breathless hush. They held their breaths, unmoving, that they not disturb her presence. She floated up the steps in the sanctuary, the light about her shimmering and growing as she rose. At the level atop the stairs she paused and looked quietly at the deep-set eyes of the holy man waiting there. He spread his arms wide, slowly welcoming her presence on the dais. Soft sighing passed through the great room and upward through the vaulted ceiling, rolling back and forth until it faded away once again into silence. Infants made no sound, round eyes peering somewhere, seeing something perhaps known only to them.

All focused in expectation on the girl child now perfectly still. The oscillating aura surrounding her began to change in appearance, first to pale peach, expanding to obscure the old man in front of her, then slowly pulsing to lavender. She turned to face the throng and as she did so she raised her arms toward them. Her palms emanated a forest green misty light of their own. She stopped, now facing the throng and the children in front sitting in chairs with wheels and tended by caring adults.

Tears of love streamed down the cheeks of the waiting people in the huge room. Audible sobs punctuated the stillness. The eyes of the girl focused somewhere, somewhere very far in the distance. A light arose within them and shone forth, azure blue, unblinking and alive with its own fire. A warm wind arose caressing the crowd gently as it rustled her garment into slow billows of soft folds. It brushed the flaxen hair from her face and left it floating about her head as a gentle halo.

The wind appeared to construct large beads around her neck, a series of soft orbs of the same forest green light emanating from her hands. When complete, a slight crackle, crisp, unseen, could be heard. From the great organ in the cave cathedral, a deep rich sound arose, perhaps the strange wind finding its way into the huge pipes.

She now began to knead the orbs of light bringing them alternately close, palms slightly separated, never quite touching, then returning them to shoulder-height. As this motion was repeated over and over, the forest green light expanded and permeated the cave. All could feel a change, and they were in it. A great rich *om* swelled within the cave cathedral. It rose from the human mass, from the very walls, from the nooks and crannies of the walls, from the wooden benches themselves. More great sobs of freedom arose from the people thus enshrined and in various stages of change, for all understood the weight of the moment, of the one Universal God of all Peoples.

As the great *om* swelled, cries and sounds of relief echoed through the crowd. No race, no religion, no differences showed. All eyes were riveted on the movement and the light of the girl. Screams of release punctuated the sound from the crowd. She rose slowly, an inch, two or more, lifted by the growing mass of light now heavy with its presence, growing, reaching out to the chairs with wheels and their small occupants. And the great *om* deepened, resonating from every form and structure, every person in the cave, the very walls of the cave in the mountain. Above and through the bedlam in the cave the voice of the girl was heard. Though her lips did not move, her vision of blue light, now piercing and vast, did not waver. Her voice rang clear through it all flooding the great cave of the cathedral. Her words spilled out of the cave, powered by the myriad minds of the New Earth, those minds to whom the challenge fell to keep what was good and to rebuild the New Human. Her words, clear and bell-like, echoed around the world, a universal message: "Let there be healing! God is here and God is every one of us!"

The wind changed to a chorus of sound, a sound of connectedness, oneness, beauty and power, a sound of union and togetherness. Then the resonance subsided and the chorus continued, now softly. The green light blended into a quiet lavender and the girl stood still, hands at her sides as soft cries of thanks poured forth from those present, and the children rose to greet her. All knew oneness and a view of humankind's new path and concordance.

Jason and Greg stood on a hillock at the edge of the valley. In front of them now these many months later, plants stretched as far as they could see, green and yellow, red fruited and fragrant, carpeting the valley. The plants whispered to each other beckoning them to approach. They stepped forward and were overwhelmed with a sense of salvation and awe. The plants had grown tall and healthy from the greenish-purple ground cover that had appeared only a few months before. At first they thought they might be giant sunflowers, but as they approached they realized they were something new, something they would not find in any botanical books.

Recent months had brought horrendous changes, both in the world and within them personally. Long before the great event they'd worked hard to establish food and emergency supplies for their city but they'd had no way to provide for permanent sources of food. Despair and anger had warred within them at their helplessness. Now as they surveyed the lush and bountiful scene before them, tears of relief and joy misted their eyes.

The two men heard it at the same time. With one voice it spoke to both.

Father! Father! See me. Hear me. I am here. We are together. Take my hand, father, and we can go far into the future and still be here in this place. Come, count with me, one, two, three, four . . ."

The two men counted and a wind arose, blowing from somewhere, pushing them, yet neither disturbing the sand upon which they stood nor fluffing the long white robes worn by the people. Their counting was overtaken by some other force, for they did not recall speaking any words, and as the count progressed its pace picked up until the words ran together into the sound of a rich chorus of voices breathing into a great *om*, resonating through their beings.

Abruptly it ceased and they were standing on a small hillock in a place that appeared as the one where they started and yet different. There were fewer animals, different plants and a great indentation in the sand in which were gathered many people, standing quietly, most clad in white flowing robes.

They faced east, quiet and still as if waiting or listening to something. Those furthest from them were kneeling in personal prayer.

Flowers, enormous swirls of light roiling in rhythm with the swelling other-earthly tones, now bright fierce orange flowers, now soft quiet pastels, each filling the heavens with its own song, all songs summed into a great sound, reverberating through the flesh and bones of the onlookers, rapt at the beauty and magnificence of the display as they dissolved into light and tone and rose softly one by one, ecstatic and free.

XXI — ONWARD

The wee blue planet, now resting from its difficult birth, traveled slowly around its Sun to which it clung, floating together through an endless universe festooned with countless tiny worlds.

Mahrianne turned toward the star-filled horizon. As one, the eyes of the Guardians followed hers. They waited. The still air gave no wind to the setting, no blade moved in its dance. Then from everywhere, from nowhere, a faint trilling sound blew about them, now here, now there. From the point at which they peered a breeze of another dimension fluffed their garments and rustled their hair.

Abruptly a low echoing tone welled from the horizon, soft at first. Iridescent and undulating, the tones swelled and built, new tones joined others until a mountain of pulsing, throbbing energy beat about them. The sound could have been dimensionless didgeridoos joining in a symphony of resonance that penetrated and surrounded their souls. Through it could be heard the tinkling of bell-like tones, celestial wind chimes responding to the winds of space-time.

An eye in the space-time continuum opened and grew. Extending her arms slightly Mahrianne faced the eye, turning her palms up, collectors of knowledge pouring over her from another dimension. A rich voice no louder than a breath cast on a wave, spoke words for the minds collected there to carry out their task. *Mahrianne, your task is complete. There is yet work to be done elsewhere. You return no more. Ascend, Mahrianne, ascend. Step into the eye open before you. Do not tarry!*

Softly, barely audible, *Come home, Mahrianne, we await you.*

She raised her hands to her shoulders and loosened the simple ties binding her robe. Time warped by the massive energy of the eye, her robe drifted in slow motion to the ground where it turned to dust, sparkling as it did so.

She raised her bare arms, earthly flesh bathed in the lights and vibrations emanating from the eye. She stepped forward, her body accelerating and becoming an iridescent light of all colors and hues. A sound more divine than earthly grew from the trilling of chimes into an all-consuming symphony of sounds, sounds from other than earthly dimensions.

Her light body passed into the eye, disappearing as the opening abruptly closed. With it the sound trailed away, echoing dimmer and dimmer, silence and stillness suffocating the hilltop as if all air had been sucked into the eye as it closed.

The Guardians faced each other and joined hands. With their eyes closed the circle began to move first clockwise then counterclockwise. Softly, then building, they joined voices in an ancient chant. The bell-like trilling came and went, came again and then faded, repeating like an echo in a vast canyon, at last fading away altogether. With each sound, a rune tumbled onto the ground within the circle of Guardians. One by one the Guardians retrieved the runes and placed them onto the hilltop altar until the last rune fell from the air and found its place for all time.

About the Author

Author Nancy Shaffron has been a life-long student of the many religions and cultures whose stories foresee the end of the world. Shaffron combines these stories with a strong background in mathematics, physics and computer science to weave a compelling tale of the end of our current world and the beginning of the final Fifth World in which all races live in harmony.

www.ingramcontent.com/pod-product-compliance
Lightning Source LLC
Chambersburg PA
CBHW061252280526
45784CB00002B/743